Intermediate GNVQ Construction and the Built Environment

Intermediate GNVQ
Construction and the
Built Environment

Second Edition

Des Millward
Kemal Ahmet
Jeff Attfield

LONGMAN

Addison Wesley Longman Limited
Edinburgh Gate, Harlow
Essex CM20 2JE, England
and Associated Companies throughout the world

First published 1998

British Library Cataloguing-in-Publication Data
A catalogue record for this book is available from the British Library.

ISBN 0-582-31560-3

Set by 32 in $9\frac{1}{2}/11$ Sabon and News Gothic
Printed in Great Britain by Henry Ling Ltd, at the Dorset Press, Dorchester, Dorset.

Contents

Preface

Qualifications for technician occupations associated with construction are undergoing development by the National Council for Vocational Qualifications. This has resulted in a qualification known as the General National Vocational Qualification: Construction and the Built Environment. The General National Vocational Qualification (GNVQ) replaces the former BTEC First Diploma and National Diploma awards, and is available at Foundation, Intermediate and Advanced levels.

The GNVQ qualifications are awarded by the City & Guilds of London Institute, the Royal Society of Arts (RSA) and EdExcel (formerly the Business and Technology Education Council). GNVQs consist of mandatory, key skills, optional and additional units; the syllabi for the mandatory and key skill units are common to all three awarding bodies. This textbook has been designed to meet the revised requirements of the three bodies for the four mandatory units of the Intermediate award.

While the text has been primarily designed to satisfy the requirements of the Intermediate GNVQ, it will also be useful reference for the Foundation and Advanced awards, plus the relevant optional and additional units, and National Vocational Qualifications (NVQs).

The book has been written in units which follow the order and titles of the four mandatory units of the Intermediate award. These are:

- Built Environment and the Community
- The Science of Materials and their Applications
- Construction Technology and Design
- Construction Operations

Each **unit** is divided into sections which reflect the unit's learning requirements; key words and phrases related to these are highlighted throughout the book. Students should be encouraged by their tutors to recognise the inter-relationship of the units in order to develop an integrated understanding of all aspects of the built environment.

Spread throughout each unit are **self-assessment tasks** designed to encourage the reader to reinforce their learning, some of which integrate the key skills of Communication, Application of Number and Information Technology.

The presentation of the information in the individual units has not always slavishly followed the exact order of the unit specifications. This has been done to allow the text to be presented within a more logical structure and allow relevant and complementing issues to be placed together for the reader's benefit. It is hoped, however, that the careful organisation and sign-posting of information, together with the comprehensive index, will allow students to easily satisfy the requirements of each unit.

The text will be useful if not essential reading for those who are intending to pursue careers in building management, building surveying, architectural design, planning, civil engineering and building services engineering. It may also prove to be a source of reference to current practitioners.

The book is not meant to be an exemplar of construction and associated information, but rather to indicate the basic approaches which may be adopted to fulfil the various requirements. Where dimensions are stated, they are intended to give an idea of scale rather than be prescriptive in meeting particular requirements.

The authors, in writing this book, have been conscious of the need to reflect the philosophy of the GNVQ awards and realise that there will be important omissions apparent to the informed reader. These omissions have been made in order to reduce any possible confusion in the student due to the inclusion of material not required by the various syllabi. The authors and publishers, however, would be pleased to receive any constructive comments or suggestions that may be incorporated into future revisions.

Acknowledgements

The authors would like to thank the following for their help during the preparation of this book:

- Turhan Ahmet, Tarkan Ahmet and Richard Tomlin for preparing drawings in Unit 2.

The publishers are grateful to the following for permission to reproduce copyright material:

- Addison Wesley Longman Ltd for our Figs 1.71–1.73 from *Building Organisation and Procedures* by Forster, our Fig. 2.37 from *Materials in Construction* by Taylor, our Fig. 2.43 from *Understanding Buildings* by Reid, our Fig. 2.49 from *Materials* by Everett, and figures which originally appeared in *Advanced GNVQ Construction and the Built Environment* by Millward *et al.*
- Butterworth-Heinemann for our Fig. 2.44 from *Newnes Construction Materials Pocket Book* by Doran.
- Charles Hardway, Marketing Manager of Stocksigns Ltd of Redhill, Surrey, for the selection of their signs reproduced in Unit 4.

While every effort has been made to contact copyright holders, in some cases this has not been possible and we would like to take this opportunity to apologise to anyone whose rights we may have unwittingly infringed.

Built Environment and the Community

Des Millward

The built environment has evolved, and continues to evolve, over time. The buildings which have survived could be said to be those which were the result of a combination of successful design and the correct selection of materials. This combination was used in such a way that the performance requirements, coupled with the technological knowledge at the time, more than proved the adequacy of the choice.

Knowledge of materials was gained empirically. The reliance on trial and error taught our forebears some hard lessons. This did however motivate them to explore other possibilities in the use of materials and their use in various structural forms. Experimentation with the production of materials led eventually to advancements which often signalled the ability to employ materials in a different way, which then modified the structural form.

The arrangement of buildings in a particular locality allows us to define them as hamlets, villages, towns and cities. While an arbitrary classification of them could be one of size, other factors such as economic activity and the social needs of the community cannot be ignored.

There are many examples of these types of communities which evolved initially in an apparently unplanned way, compared to today's planning practice. Further examination will show that in fact their layout, while being influenced by the points above, will have been influenced by their geographical location. Students in studying this area will have their interest awakened and developed by being introduced to past and present planning practice.

The creation and maintenance of the built environment involves a wide variety of activities performed by a range of people. Each activity demands a mixture of experience, knowledge and training. This can lead to individuals possessing a recognised qualification in their area of work.

Further development is possible, and this can be achieved by individuals developing roles in an area of work through career progression, coupled with additional academic and professional recognition. In the future, individuals will increasingly be involved with different activities and job functions through their working life. Studies in this unit will show the range of employment possibilities in the built environment, together with associated progression routes.

1.1 The development of the built environment

Roads

Prior to the Roman occupation of Britain the landscape was uncivilised and uncoordinated. There was very little communication between the settlements until the Roman road building created links between the towns they established.

The road network spread throughout the Empire joining the major cities to one another. Initially this provided the means for the efficient movement of troops and their supplies. The Roman road network spread itself out from London rather like the 'A' roads or motorways of today. The Roman road was a series of straight alignments ignoring all but the most awkward obstacles.

Watling Street is one of the most well-known Roman roads. The Roman method of construction was to strip the topsoil and create a foundation of gravel or small stones. On top of this were large stones bedded in mortar, with a further layer of gravel and mortar on top of which were polygonal stones in an interlocking pattern. (Fig. 1.1) Following the departure of the Romans from Britain their roads fell into disrepair.

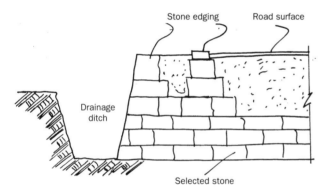

Figure 1.1 Section of a Roman road

With the exception of a few roads of strategic importance road building and repair was a matter of local concern. Tolls were levied on road users to cover the cost of the repairs. This however was inadequate for the increasing amount of traffic using the roads. A statute of 1552 made each parish responsible for repairing its own roads, with each parishioner spending four days a year in the duty of repairs under the supervision of a surveyor.

This system was inefficient as repairs were undertaken by a pressed workforce with little knowledge of current techniques. There was also little chance of developing a uniform and coherent network of roads between the parish boundaries. Strains on the road system developed as the long distance traffic increased.

The answer to the problem was the setting up of Turnpike Trusts, the first one being in 1663. The Trusts were to be responsible for the repair and maintenance of ten to twenty miles of road. They had the right to charge tolls to support road maintenance costs. The Trusts improved the roads by using good construction methods developed by the great road

engineers. Travel by coach throughout the country was much improved.

Though there was no national plan to the development of the turnpike system the roads which were turnpiked were those which were the most heavily used, especially those into London before 1750. By 1770, 15 000 miles of roads had been gated and tolled. The general development of the economy led to turnpikes in areas away from London which served the growing provincial towns and cities. Three hundred and eighty-nine new trusts had been established in the period 1750–1772, which started to lay the basis for a coherent network of roads.

Metcalf, a Yorkshireman, discovered that angular broken stones made a better road foundation than smooth round ones which were pushed aside when placed under load. He subsequently supervised the construction of 180 miles of roads in Yorkshire built this way. New roads and better coach construction encouraged more travel.

Macadam in 1804 designed a road which was built by first stripping the topsoil and grading a formation to required cambers to assist the road in draining to a ditch. Above the formation was a layer of broken stones which were not blinded with smaller stones but left to traffic to consolidate and fill the voids (Fig. 1.2).

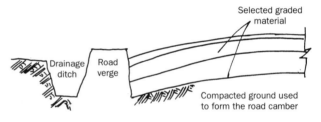

Figure 1.2 A Macadam road

Telford, another engineer, created a road camber in the material used for the formation level. This was filled with smaller stones which assisted in bonding the surface together. On top of this there was a layer of 50 mm broken stone, which was topped by a further layer of smaller graded stone which provided the riding surface (Fig. 1.3).

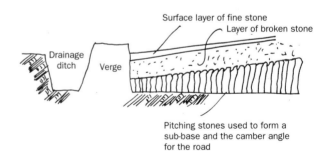

Figure 1.3 A Telford road

In towns the roads were paved with stones, where it was found that a flat topped stone would give better riding qualities. These stones were known as **setts** and were often made from granite. Wooden setts were also used.

Asphalt was introduced in 1836 but not used for the surface of carriageways until 1870. Towards the end of the nineteenth century the motor car was appearing and it soon became apparent that the roads would need to be constructed differently to produce a smoother ride. Road rollers appeared which were horse-drawn at first but later became mechanically propelled. This allowed the rollers to increase in weight and to produce better consolidated roads.

The arrival of the canal network captured much of the heavy freight from the roads, and railways took away the long distance passenger traffic. Roads became only local feeders to the nearest station and from the 1840s the Trusts largely became redundant. By 1895 the whole turnpike system was wound up, and the care and maintenance of the roads was given to the emergent local and county councils.

Horse-drawn tramways were introduced in Britain in 1861 but the rails which stuck above the road level were disliked and the line was closed. The use of steam brought trams back into favour, but their most successful period came with the introduction of electrically propelled trams in 1885 in Blackpool. By 1914 electric trams had virtually put horse-drawn traps off the road, the trams providing the cheapest ever form of mass urban transport.

Motor cars in Britain were still a rarity in 1901, roads being mainly used for horse-drawn vehicles. Roads were rolled macadam in the country and the suburbs, with stone setts in the industrial towns. In 1909 serious consideration was given to adapting the roads to the needs of motor vehicles. A road board was set up which raised duty from cars according to their horsepower. In 1919 the Ministry of Transport took over from the road board and there followed road widening and by-pass schemes to assist the unemployed.

In 1935 the Restriction of Ribbon Development Act controlled the spread of building construction alongside the main road arteries. In 1949 the Special Roads Act allowed building of roads purely for the motor vehicle. This was the go-ahead for motorways in Britain. Up until this time the road network had consisted mainly of main roads ('A' roads) and subsidiary roads ('B' roads). The country's first motorway, from London to Birmingham, was opened in 1959. It was capable of carrying 60 000 vehicles a day on three-lane carriageways. Since that time the building of motorways and other trunk roads has continued apace.

Many of the early road-building principles are still applied today, namely:

- drainage
- a running surface
- a foundation or sub-base
- horizontal and vertical alignment

Two principal forms of road construction are used today, these being:

- rigid
- flexible

Rigid pavements consist of the sub-grade to which is added the sub-grade on which the surface slab is laid. Rigid pavements have the ability to accommodate movement and yet remain stable. Construction details are shown in Unit 3.

Flexible pavements are able to maintain their structural integrity when small vertical movements take place at the surface. The pavement consists of surfacing course, road base and sub-base. Again further details are in Unit 3.

Canals

As the Turnpike Roads were being built in the reign of George III, England was being covered by a network of canals. The industrial revolution meant that cheap transport was needed for carrying raw materials and finished goods. The roads were considered to be too bad and expensive as a regular means of transport. Rivers, while useful for moving large and bulky cargoes were unable to reach the growing number of industrial areas which did not have navigable rivers serving them.

In 1755 the first Canal Act was passed which allowed the building of the Sankey Canal from the Wigan coal field to the River Weaver. The rapid spread and use of canals was due to the Duke of Bridgewater who obtained an Act of Parliament in 1762 to build a canal from his estate at Worsley to Manchester. This carried coal from his own coal field and effectively halved the cost of coal.

James Brindley was the most important canal engineer of the time, having worked on the Leeds, Derby, Birmingham, Salisbury and Glasgow canals. Water transport and coal linked together to form an important part of the industrial revolution. The canal building commenced in the 1760s continued at a faster pace, twenty new canals being approved by Parliament in 1793 alone. Not all canals which were planned were actually built, but 4000 miles of canals were in existence by 1830. Then the opening of the Liverpool and Manchester Railway began a new transport revolution. Canals lost their monopoly, and were forced to lower their prices to compete with the railways. However, by 1840 the use of canals was declining. A few remain for commercial use, and some are navigable by pleasure craft.

Railways

The railways provided a faster, cheaper and more reliable form of transport than offered by roads or canals. In 1821 George Stephenson was asked to plan the Stockton to Darlington railway to carry coal. This eventually opened in 1825, and is usually taken as the start of the 'railway age'.

In 1830 the Liverpool and Manchester railway opened after having overcome many construction difficulties. It was to carry passengers and goods such as cotton. Early railways met local needs only but eventually cross-country routes were built such as the London to Birmingham line completed in 1838.

Railways began to be built elsewhere, unfortunately to different gauges, which meant travellers had to change trains to complete journeys. In order to resolve this problem the government made the narrow gauge the standard width for the whole railway system, though the broad gauge continued to 1892 when it was abandoned.

The railways had very important social and economic results. They carried raw materials and finished products cheaply and quickly. Railways connected virtually every major town of importance, such was the scale of their importance to industry.

Expansion continued but eventually the decline of some industries and rationalisation of the railways by Dr Beeching, a government minister, led to the closure of many of the branch lines and the need for the population to consider other forms of goods and passenger transport. This caused the gradual switch to road transport, and the requirement to build more roads to higher standards.

Architectural styles and features of different periods

Britain possesses a rich architectural heritage which reminds us of our predecessors, the life they lived and the buildings they constructed. It is a tangible record of the materials and construction techniques which served to meet the needs of all members of society over a very long period of time. The designs which were used have been repeated by others, albeit perhaps by using contemporary materials and techniques but aiming to make good use of best practice from the past. It is only natural therefore that study of the built environment should form a part of a programme of studies.

Table 1.1 General chronology

Date	Period	Style
1066–1558	**Medieval**	
1066–1200		Anglo-Norman
1200–1300		Early English
1300–1400		Decorated
1400–1500		Perpendicular
1500–1558		Tudor
1558–1600	**Transition**	**Elizabethan**
1600–1702	**Renaissance**	
1603–1660		Jacobean
1660–1702		Restoration
1702–1837	**Georgian**	
1702–1760		Early
1760–1805		Late
1805–1837		Regency
1837–1910	**Revivals**	
1837–1901		Victorian
1901–1910		Edwardian
1910		**Modern**

Norman

Norman architecture is represented by the castles built by the aristocracy as fortified homes to control the Saxons. Churches and cathedrals were built by the Benedictine and Cistercian monks. Over two hundred castles were built in the Norman period, these being built in every shire and town occupied by the Normans (Fig. 1.4).

Figure 1.4 Norwich Castle built from Caen stone

The Normans were essentially builders in stone, producing buildings of massive construction, the walls being so thick that buttresses were minimal in size (Fig. 1.5). Mouldings round doorways were ornamented with geometric designs. Arches over windows, doors and vaults were semi-circular. Supporting columns were massive with simple capitals (Fig. 1.6). The vast quantity of stone required by the Normans meant that imports of limestone from Caen were necessary. Roofs were made from timber trusses which were boarded to form a ceiling finish.

Lords of the manor lived in manor houses based on cruck construction, or stone-built manor houses. The serfs continued to live in wattle and daub cotes based around the castles and manor houses.

Early English – Gothic

Castles continued to be built and monks vied with friars for cathedral building. Additions were made to castles such as kitchens, butteries and chapels. Gothic architecture developed, where the chief principle was the concentration of the weight of the building on isolated points by means of ribbed vaults, the pointed arch and the buttress (Fig. 1.7).

The introduction of the Gothic arch created a point which removed the limitation of the semi-circular Norman arch. Arches could be made more pointed than each other, which

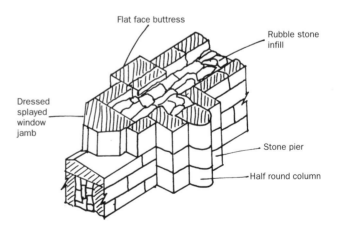

Figure 1.5 Norman wall and window detail

Figure 1.6 Norman column

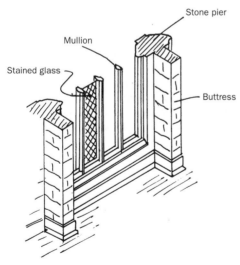

Figure 1.7 Gothic wall and window detail

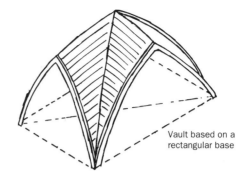

Figure 1.8 Gothic vault

Figure 1.9 St Mary and All Saints: 'Crooked Spire', Chesterfield

of columns. Windows became Lancet-shaped and narrow, so they were grouped together under one arch making a wide mullioned window (Fig. 1.9). Spires and steep roofs assisted in making buildings lighter. Roofs were formed using hammer beam roofs (Fig. 1.10).

Decorated – Gothic

The country's prosperity manifested itself in the cathedrals and abbeys as well as in the parish churches. The architecture broke away from its previous severity and became highly ornate, which gave rise to the name 'decorated' for the style. Stonemasons could construct thinner structural members and they could be placed further apart. Flowing tracery was introduced to windows (Fig. 1.11). Vaults became more elaborate with extra ribs called 'liernes' forming web-like patterns. The junctions of ribs were often decorated with carved bosses.

Manor houses had open timber roofs and an opening to let

allowed the space being vaulted to be rectangular rather than square (Fig. 1.8).

Windows became larger which required the walls to be stronger, and so the buttresses in the walls became deeper. Norman columns were replaced by piers consisting of clusters

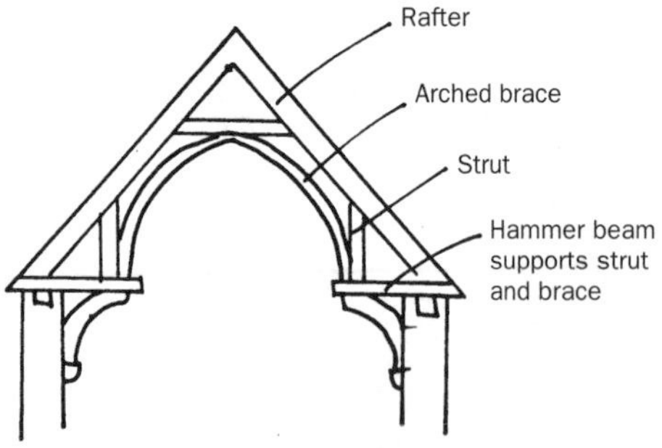

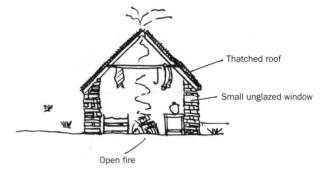

Figure 1.10 Hammer beam roof

Figure 1.12 Cottage detail

Figure 1.11 Decorated tracery window

out the smoke from the fire. Walls were covered with plaster, wood boarding or tapestries. Floors were of clay with rushes spread over them. Cottages of the people were mainly single storey constructed of mud and wattle, wood or stone with thatched roofs and small windows (Fig. 1.12). During this period the Black Death caused many fatalities which resulted in a slowing down in building work in the latter part of the fourteenth century.

Perpendicular – Gothic

The intense activity of earlier periods and the large number of churches which had been constructed, coupled with the Black Death meant little major new building work was undertaken. Any work which was undertaken was due to the modification or enlargement of existing buildings.

A more severe perpendicular style superseded the decorated, being the last stage in the evolution of Gothic architecture. Stone construction was carried to its limit, allowing the buildings to become light, open structures where the walls were reduced to rows of thin piers between wide

windows. Light flying buttresses were used to control the weight of the building together with the outward thrust of the vaults. Externally roofs became virtually flat with high pierced parapets relieving the monotony being silhouetted against the roof.

The English squire started to have the bedroom at first floor level and the cottage took on a more rectangular plan. The walls were single storey stone and rubble faced and topped with a sand, clay and straw mixture. Timber posts and lintels were used around door and window frame openings. The roof was thatch on a light timber frame construction. Chimneys and fireplaces started to be placed at one end of the building.

Peasants' houses changed with the development of the new-found middle class. Stone gables started to become more common, assisting with the stability of the structure. Roof framework was still visible from the inside, with bracing starting to become more ornate. Square and rectangular windows appeared in relatively long lengths which assisted in admitting more light.

Tudor

Church building declined following the dissolution of the monasteries. The wealthy commissioned the building of palaces and mansions. This was the start of the transition period where the architecture was Gothic in form but altered and adapted to suit the domestic properties.

Walls carrying the structural load were interrupted by large windows constructed from stone mullions and transoms (Fig. 1.13). The horizontal transoms together with a flattened arch destroyed the Gothic verticality. The opening portion of the window was made from iron and hung from hinges on the side. Limitations were placed on the size of windows which influenced the spacing of the mullions, thus forming small rectangles which were a feature of sixteenth-century architecture. Vaulting and tracery started to disappear.

Houses were timber framed, being constructed on the box-frame principle (Fig. 1.14). The panels formed between the frames were filled with lath and plaster or in some localities with local stone. Brickwork started to become more popular and was used in the houses of the more wealthy to form decorative patterns.

The building of mansions included a greater number of rooms and the inclusion of brick-built fireplaces. The chimney stack became important and was a prominent feature of Tudor timber frame houses. Greater consideration was being given to interior comfort with the ceilings being plastered and painted, and walls panelled with wood or hung with tapestries.

Italian renaissance started to influence the design of

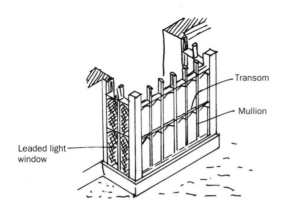

Figure 1.13 Tudor wall and window

Figure 1.15 Hardwick Hall, Derbyshire

Figure 1.14 Tudor house and shop (Lavenham, Suffolk)

Figure 1.16 Classic wall and window detail

mansions in the Elizabethan period. Building plans became more balanced and symmetrical using bay windows and towers (Fig. 1.15). Windows became exceptionally large with square heads, mullions and transoms, with no sign of Gothic tracery to be seen. Interiors were elaborate decorated plaster ceilings and oak or plaster panelling.

Homes of the middle classes were being re-built in brick. Glass was becoming cheaper and came into everyday use. Walls were panelled and ceilings plastered, with floors becoming carpeted. These transitional styles carried on into the seventeenth century and came to be called Jacobean.

Jacobean – Classic

The Gothic building arose from the exploitation and development of particular materials. Classic buildings were constructed to pre-conceived designs where the construction and materials were secondary.

The wall of Classic architecture carried the weight of the wall and was unbuttressed (Fig. 1.16). This served to limit the size of the window which resembled a hole piercing through the wall. Wooden windows were used with transoms being placed two-thirds up the height. Opening sashes were glazed using leaded lights.

Large buildings were designed in the style of Inigo Jones who followed the Italian designer Palladio, which led to Jones' work being referred to as Palladian. Exteriors were more symmetrical with roofs being hipped and sometimes being hidden by parapets (Fig. 1.17). Internally plaster, marble and paint came to be increasingly used and replaced the traditional material of oak.

Humbler houses continued to be closer to the tradition of the preceding periods, being more similar to the perpendicular period of the Gothic with sparse detailing. Local materials still dictated the character of the structure, with thick walls of hard millstone grit and low pitched roofs of heavy grey slates. In other areas roofs were covered with stone slabs in diminishing sizes. Windows were small metal casements set in deep stone mullions.

Figure 1.17 Blickling Hall, Norfolk: Jacobean mansion of 1626. Main entrance with Doric columns and entablature. Note also the use of lead rainwater pipes and decorative hopper heads

Georgian

Building activity in this period centred around domestic, college and ecclesiastical buildings. The wealthy vied with each other for country mansions, and trade expansion and government departments led to Town Halls, Corn Exchanges and Customs Houses. Larger buildings reflected the influence of Inigo Jones and Christopher Wren.

Early Georgian buildings saw improvements in construction techniques and the almost universal use of brick which allowed walls to be thinner. Coloured bricks were used to emphasise the window openings, replacing the previous period's decoration of architraves and cornices. Sliding sash windows came into general use which made use of the larger and better quality sheets of glass being produced. Panelled doors were designed to minimise the movement of large pieces of timber and were emphasised by mouldings in the frame.

The buildings of the Late Georgian period were very much of the same type as the start of the period but with fewer mansions and churches being erected. The earlier style becomes refined and simplified as the expertise of craftsmen and designers improved. The panels of doors and wood-based wall coverings became larger and the mouldings more refined. Glazing bars in windows became lighter and the box frames of sash windows were concealed in the brick walls (Fig. 1.18).

Heavy cornices became replaced by stone bands or slight projections. Also receiving attention were wrought iron

Figure 1.18 A restored urban Georgian terrace built 1761–71, Surrey Street, Norwich. Note the bay, dormer and inset box-frame windows

railings, lamp standards and staircase balustrades. Robert and James Adam influenced design trends, being known for their elegance and lightness. They favoured flat, elliptical or curved arches, and circular or oval ends to rooms. Wall decorations consisted of semi-circular niches or circular plaques.

The classical styles and interior designs of the Georgian era continued into what was known as the Regency period, so named after the Prince Regent who used John Nash to turn London into a planned city. Nash assisted in the rebuilding of Buckingham Palace and designed the Marble Arch in London. The idea of the crescent became fashionable (Fig. 1.19).

Exteriors of buildings were finished with painted stucco, a hard plaster finish, and flat segmental bay windows. Slates were used on roofs which achieved even lower pitches than previously. Balconies were supported on slender columns of iron or stone (Fig. 1.20).

Houses for industrial workers consisted of nothing more than long terraced rows with no sanitary conveniences, and a minimum of light or air.

Revivals – Victorian

This period saw a huge rise in the numbers of buildings being constructed, mainly due to a rapidly increasing population in

Figure 1.19 The Crescent at Buxton, Derbyshire, completed 1790

Figure 1.20 Balcony supported on cast iron brackets

Figure 1.21 Norwich Union building in Edwardian style

Figure 1.22 Norwich railway station

Figure 1.23 Iron trusses at Liverpool Street Station, London

the towns and cities. There was no consistent architectural character which led to designers reviving past styles, of which the Early English Gothic was the most popular. This was chiefly employed for houses and churches, and the Classical style was used for civic and public buildings (Fig. 1.21).

Improvements in communications caused the new architectural styles in railway stations (Fig. 1.22), bridges and engine sheds. The obvious characteristic was the use of glass and iron for big span halls and bridges. Stations and halls had iron arches or trusses and columns to carry the roof weight down to isolated points (Fig. 1.23). Sheets of glass were fixed to wood or later metal glazing bars.

Lattice construction for trusses and arches contributed to the structural rigidity of the frames, particularly when based on triangulation principles. Repetition of standardised units

and cross-sections became available as manufacturing techniques for steel improved.

Typical houses of this period for the workers were terraced two or three storey properties with a basement. Slate roofs, sash windows and stained lead lights were also key features. In addition there were fireplaces in every room and outside sanitary facilities. Back-to-back houses were also common in industrial areas where the inhabitants would share a communal tap and toilet facility. These were to become known as slums.

Modern

Large numbers of civic, commercial and ecclesiastical buildings were erected. New building types such as swimming pools and cinemas made their appearance. Positive housing policies led to four million houses being constructed after 1945 and blocks of flats replacing the slums.

Advances in architectural education brought about better planning and academically correct buildings, which tended to reflect the earlier Wren and Georgian styles. The Classic proportions were still employed, and walls having the appearance of being load-bearing were in fact hiding a structural steel frame.

Self-assessment tasks

1. Using sketches describe how window openings developed between 1066 and 1500.
2. Explain with the aid of sketches how the use of the segmental arch in Gothic architecture helped to increase the span between supports.
3. Describe how the development of materials was reflected in Victorian architecture.
4. Sketch details to show how the stone columns changed in shape in the Norman and Gothic periods. State reasons why this change occurred.
5. Describe how the Georgian period influenced contemporary buildings.

Figure 1.24 Typical Guild Hall, Thaxted

Town planning

Prior to the Industrial Revolution society was reasonably self-sufficient. In feudal villages those whose primary concern was not farming would have possessed a specialist trade such as carpenter, blacksmith, miller or weaver. Often these trades would have been carried out part time alongside agricultural duties from the family dwelling house.

Defining the difference between a town and a village was virtually impossible. The existence of a successful market is probably the best guide. Towns were places where a significant amount of trade would have taken place, and where there would be a greater number of full-time specialists across a wider spread of trades. Markets provided the ideal outlet for the sale and purchase of surplus produce, together with goods not normally available in the surrounding villages.

Markets were usually held weekly, and occasionally more often, using stalls set up for the day, not unlike some markets held in present times. The introduction of a new market could only be by royal consent, and often relied upon there being no other market in close proximity. Thus the development of trading centres and their limitations were born.

The medieval markets competed with each other, with the most successful being located on well-established transportation routes, either by land or water. The most notable markets were Dunstable, Stony Stratford and Atherstone on the Roman highway of Watling Street. Development and location of the buildings would have been generally unplanned by today's standards, but were based around the church and the market area. Goods would be brought to the market from a wide area, and the more successful market towns would have markets on different days for different goods, such as:

- meat
- fish
- livestock
- grain

Corporate status was conferred on cities and towns around 1200 which allowed varying degrees of self-government and the development of local administration centres. London, which was to become the most important and powerful, was the first city to attain such status. Corporations came into being, often being connected to local guilds and merchants.

Medieval development

The typical medieval town was one that expanded naturally because there was a reason to do so. Often an excellent trading location combined with skilled craftsmen producing good quality goods enabled experienced merchants to place them into profitable markets. Towns also could gain the privilege to hold a market. This was either held in the church or in the market square near to a market cross.

The merchants in medieval times, despite gaining greater freedom to manage their own affairs, did not comprehend the need for town planning, nor the longer term aims for the development of their community. They saw their town as a place where every building served a purpose and was therefore built for use and occupation, not for speculation. The need for additional accommodation was met when the demand arose and so the town evolved slowly over a period of time.

Medieval people placed great emphasis on the market place (Fig. 1.25) as a civic centre where they held processions, plays and public meetings. The broad market place contrasted with narrow confined corridor streets (Fig. 1.26); likewise the narrow-fronted shop or house with high guildhall (Fig. 1.24) or church. Domestic construction appeared frugal at the side of quality materials in churches and cathedrals. Sewage and water distribution systems were non-existent. Streets and pathways were inadequately cleaned, drained or paved.

The siting and development of medieval towns and villages moved away from the need to be in strong defensive positions, though siting them close to water crossings where there was a barrier to natural communication was still influential. The shape and size of towns and villages were infinitely variable but two common forms exist. These are based around:

- a street or road
- a square or triangle of land

Villages based around a street or roadside would have buildings placed unselectively. These would have consisted of houses, shops, hostelries and churches plus others as appropriate. The village may have been situated at a junction, and stretched a little way down each road (Fig. 1.27). Often the road widened prior

Figure 1.25 Norwich market place

Figure 1.26 Typical fourteenth-century street, Elm Hill, Norwich

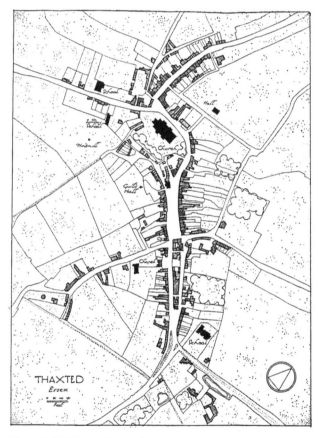

Figure 1.27 Thaxted town plan

to the market hall or guildhall (Fig. 1.24) which would be able to accommodate the stalls on market day. Figure 1.28 shows a plan of a simple street village.

'Square' villages were often anything but that. They could be rectangular or triangular in shape. The roads entering the square were often staggered and did not provide a straight road running right through the village. Views into the square are limited by the buildings around its perimeter, and views out of the square are limited by the vista provided by the roads (Fig. 1.29).

Renaissance development

The strength of the degree of control of the guilds over industry and trade gradually weakened and during the reign of Elizabeth I gave way to unrestricted production and national control over wages and prices. Specialist manufacturing areas such as Sheffield, Birmingham, Norwich and Leeds made rapid progress – the economic activity expanding at the hands of landowners, manufacturers and merchants, bringing them great wealth. London led in being the centre for overseas trade, providing banking and financial services.

London and other cities initially retained a medieval appearance and layout and were starting to become overcrowded. King Charles I wished to modernise London but lacked sufficient finance, so he set up a Buildings Commission with the intention of regulating new development. This produced the first evidence of change in planning led by Inigo Jones's church, piazza and house gardens in Covent Garden, which was the first Renaissance square.

Other developments were to follow with Lincoln's Inn, Moorfields' and Great Queen Street, said to be the 'first regular street in London'. Later came Leicester Square, Bloomsbury Square, Soho, Red Lion and St James's together with Grosvenor

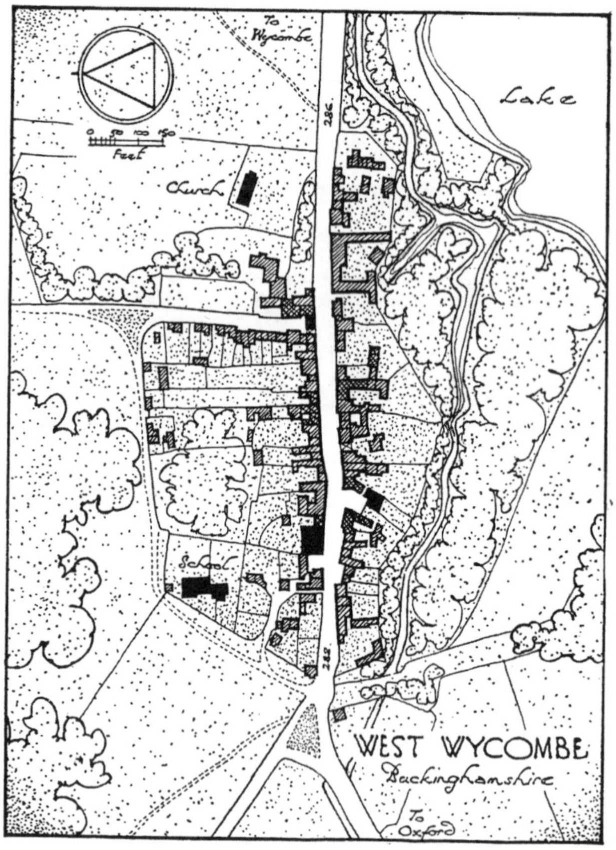

Figure 1.28 West Wycombe, Buckinghamshire

Figure 1.29 Sherston, Wiltshire

and Berkley Squares. These squares and the streets linking them to existing thoroughfares represented small-scale town planning which created significant town improvement.

The Great Fire of London in 1666 presented an opportunity to re-plan and re-develop the city. Sir Christopher Wren carried on from Jones and was given the responsibility of designing St Paul's Cathedral and many other churches. His plans for replacing existing streets with a hierarchy of avenues, streets and lanes was rejected on the grounds that it was not capable of being achieved.

The Act for Rebuilding the City of London in 1667 required the narrow medieval streets to be widened and straightened, awkward street levels to be adjusted, and bottle-necks removed. Buildings according to their importance were to be of particular heights and width and were to be built in brick or stone with tiled roofs. This resulted in urban form being controlled with regular frontages, building height and street widths and uniformly spaced windows.

Town planning in the Renaissance period concerned itself more with piecemeal additions on fields or gardens or the replacement of medieval slum areas. These were closely related to the current demand and the availability of financial backers. Sometimes the developments were extensive following serious fires, or were confined to a few streets or squares.

Georgian development

Towns in the Georgian era gained from the grace and dignity of the design of their buildings. The use of local building materials ensured that the new buildings blended with those of an earlier period, and local craftsmen, while observing the order of proportion and scale, also applied their own ideas. The Georgian façade became so popular that medieval town houses were modified to include new classical fronts and in some cases modified interiors.

Bath is an early example of Georgian town planning being transformed from a medieval walled town to a prosperous and elegant health spa. This was undertaken by Nash, Allen and Wood to produce Queen Square, the Circus (an enclosed residential space with a continuous uniform façade of approximately 100 metres diameter), the Royal Crescent (a façade built on a semi-ellipse 200 metres long, facing a grassed expanse down to the River Avon) and the North and South Parades.

Beautiful house façades lined proportioned avenues, crescents, circuses and squares interfacing with generous planting of trees, lawns and shrubs. Shop fronts were very elegant with impressive public buildings in the Palladian style. Elsewhere this age of elegance extended to and included improvements to towns such as Cheltenham, Leamington and Buxton.

Victorian development

As a result of the eighteenth-century inventions which produced efficient power-driven machinery and extended production for a wide range of manufacturing, town populations grew quickly (Fig. 1.30). Concentration of manufacturing favoured certain towns such as Manchester, Birmingham and Sheffield.

A massive shift of the population took place from the countryside to the town. Figures indicate that 200 000 people had to be rehoused but town councils or employers saw it as their role to provide the necessary dwellings. Instead 'jerry

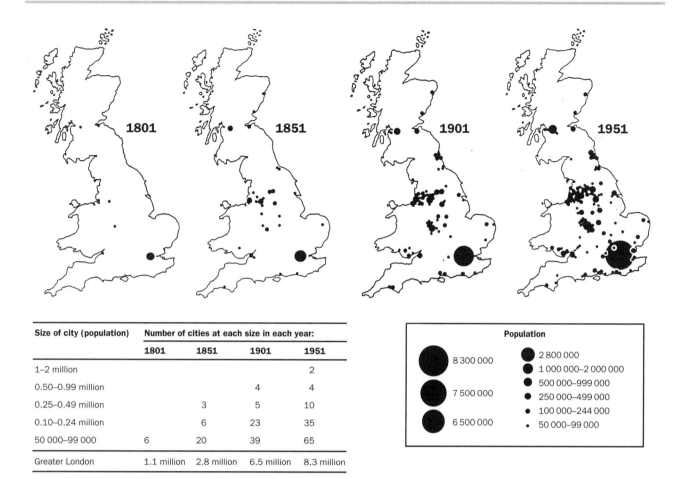

Size of city (population)	Number of cities at each size in each year:			
	1801	1851	1901	1951
1–2 million				2
0.50–0.99 million			4	4
0.25–0.49 million		3	5	10
0.10–0.24 million		6	23	35
50 000–99 000	6	20	39	65
Greater London	1.1 million	2.8 million	6.5 million	8.3 million

Population			
8 300 000		2 800 000	
7 500 000		1 000 000–2 000 000	
6 500 000		500 000–999 000	
		250 000–499 000	
		100 000–244 000	
		50 000–99 000	

Figure 1.30 Growth of cities

builders' or speculative builders undertook the task. Regrettably their aim was to maximise the use of land with minimum expenditure on materials and services.

'**Back to back**' housing in double rows allowed a housing density of 60 houses per acre. They produced conditions that were incompatible with maintaining the health of their occupants. They lacked through natural ventilation, light or sanitation facilities and had inadequate water supplies. The accommodation soon became grossly overcrowded as population exceeded provision of living accommodation.

From 1839 the development of the railways brought them into the towns, adding yet more dense smoke and fumes to the already industrially polluted atmosphere. The divide between middle and working classes became more reinforced. The more wealthy middle class moved from the towns to the surrounding countryside and formed suburbs where housing was four to eight houses to the acre, leaving their Georgian town houses to multi-occupation by poor families.

Gradually committees and commissions inquired into the sanitary state of towns, the incidence of disease and the deplorable living and working conditions for the working classes. The Municipal Reform Act of 1835 transferred the power of local administrations to those who held greater sympathy for improvement schemes. Initially individuals caused action to be taken, with Shaftesbury responsible for the Factory Act of 1844 and Chadwick for the Public Health Act of 1848. Further idealist reformers were to make contributions to the planning of towns whose significance would be seen in the twentieth century.

Schemes for community planning included 'model villages' built by industrialists for their employees. Examples include Bessbrook in Northern Ireland in 1846, Copley, Bromborough and Akroyden. Unfortunately, by 1870 the well-intentioned ideas of the philanthropists had been limited in practice to only a few villages being built. Victorian towns and cities continued to expand to unmanageable sizes, and with increased death rates from cholera.

The Public Health Act of 1848 was replaced in 1875 by one which imposed a duty on local authorities, enforcing them to provide improved sanitary conditions and empowering them to make by-laws governing town growth. Model by-laws, adopted by the majority of local authorities, dealt with street widths, structure of buildings, minimum space for air circulation and ventilation, and drainage and sanitary facilities.

Meanwhile much of the wealth of the era provided by institutions, individuals and councils was used to provide a wide assortment of social buildings. These included museums, concert halls, art galleries, hospitals and institutions, churches, town halls, commercial buildings and warehouses.

Garden Cities

Further advances in town planning were made by W.H. Lever and George Cadbury. Both these men saw that the enforced conditions of town dwellers could be relieved by building new factories in new villages containing houses with gardens and wide open spaces. Consequently Port Sunlight and Bournville were built.

Ebeneezer Howard sought to combine the best features of town and country life in a new form of urban development (Fig. 1.31) He proposed that there should be a limitation on the size

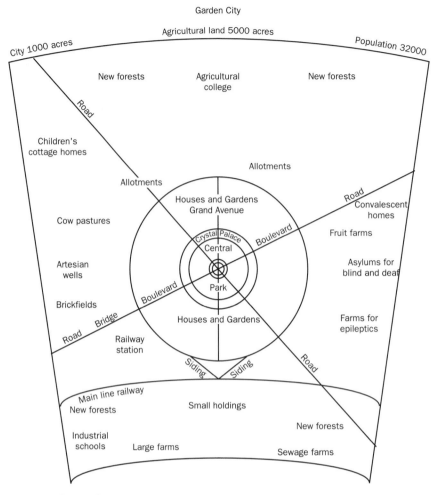

Figure 1.31 Ebeneezer Howard – Garden City

of the town, and the surrounding agricultural land should be in the same ownership. Fundamental to the idea was the fact that when the population had filled the space available another new town would be started, protected by its own green belt.

Howard planned to have functional zoning within the town. He saw commercial and industrial needs being balanced with a mixture of social groups and levels of income. Areas of activity would be worked out of the zones, and these would include:

- public buildings and places of entertainment placed centrally
- shops intermediately
- factories at the periphery with railway and sidings
- houses of different sizes, but all with gardens
- the houses within easy reach of the shops, schools, factories or cultural centres
- the provision of a central park and inner green belt

Modern development

Demand grew from a variety of areas for the introduction of legislation which would regulate town planning. Professional bodies and local administrations joined forces to press for reform. Eventually John Burns, President of the Local Government Board, introduced the first town planning legislation called 'The Housing, Town Planning, Etc. Act 1909' which aimed to 'secure, the home healthy, the house beautiful, the town pleasant, the city dignified and the suburb

salubrious'. The emphasis, however, was on raising the standards of new development.

A report, better known as the **Barlow Report**, published in 1940 was very significant as some of its recommendations were accepted as a basis for planning policy for twenty-five years. The commission advanced the idea that the development of garden cities, satellite towns and trading estates, could contribute towards reducing the growing problem of urban congestion.

The 1947 **Town and Country Planning Act** brought most development under control by making it subject to planning permission. Plans called **development plans** were to be prepared for every area in the country. These plans were to outline how an area should be developed or preserved. **New towns** were to be developed by development corporations financed by the Treasury (Fig. 1.32).

Twelve new towns were designated for England and two for Scotland. Eight of the twelve in England formed a ring around London following the Greater London Plan of 1944. All were to be complete, locally self-contained, developments of a limited population of 50 000 people and were to be surrounded by green belts. They would each have an identifiable town centre for commercial, civic, cultural and social needs serving a series of residential districts around it. In addition there would be a few areas which would accommodate factories that employed a greater part of the population (Fig. 1.33).

Integral to their development was the need to plan for motor traffic, which was forecast to increase. The danger of

Figure 1.32 Map of new towns

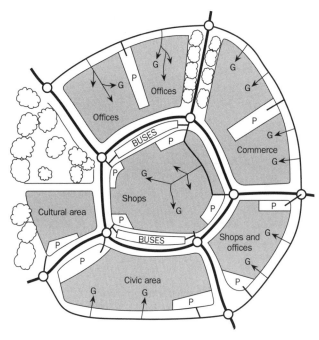

Figure 1.33 Schematic showing the precinct theory (P: parking areas; G: goods and services entries)

traffic in towns and the need to accommodate the needs of pedestrians was recognised. Shopping areas would be planned as **pedestrian precincts** where wheeled traffic access would be separated from pedestrians. Roads would be planned and identified by function such as:

- arterial
- sub-arterial
- only 'local' traffic

As towns expand it is often possible to see that different areas are clearly related to different functions. As a result, it is possible to recognise **functional zones** which are identified by distinctive types of land use. Table 1.2 summarises the chief functional zones to be found in older towns or cities.

Table 1.2 Chief functional zones in towns and cities

Functional zone	Purpose
Central business district (CBD)	An area containing the major commercial shopping, and social facilities – normally the town or city centre.
Twilight zone	A mixed area surrounding the CBD containing small industry and warehouses. Large houses have been converted for other uses, and smaller houses are old and run down. Demolition and rebuilding is a major activity.
Older industrial area	Found close to town centres when they were developing as a centre for local industry. Often built close to railways but now in a demolition area and declining (see Fig. 1.34).
New industrial area	Areas towards the periphery of the town where facilities and transportation are good. Often supported by workers who live in nearby suburbs.
Residential areas	Residential segregation occurs through grouping according to income. Will consist of new and old accommodation at low, medium and high cost.
Suburban shopping centres	Supportive shopping centres for local areas which can serve people's immediate needs.
Commuter zones	Areas outside the town or city from which workers travel to employment in the centre.

Issues relating to the development of the built environment

The continued development of the built environment may meet some of the demands of contemporary society; however, in doing so other problems may result which are not particularly palatable to members of that same society.

The increasing use of the motor vehicle has led to various concerns which are:

- pollution from the exhausts
- excessive road traffic noise levels
- inability for road networks in some towns to accommodate the volume of traffic flow or its intensity
- the need to exclude traffic from town/city centres
- the safety of pedestrians

Figure 1.34 Derelict inner city area between the river and railway in Norwich

- increasing speed at which traffic moves
- inability of bridges to carry increased traffic loads
- how to deal with or provide off-street parking

Development policies for zones of use have raised concerns over:

- out-of-town superstores
- shopping centres in towns away from the central areas
- rating policies leading to empty buildings
- vandalism of empty properties
- tracts of waste land for long periods of time
- inadequate support for renovation or conversion of prime sites
- lack of local shops in residential areas
- breaking up of established communities in areas of towns
- providing accommodation for the underprivileged

Policy measures responding to current developments

The problems described in the previous section may not all be solved by implementing policy measures imposed from central government. There is no doubt that they may assist in certain cases, but may not always prove to be the final solution.

Contemporary society is becoming more aware of environmental and heritage issues than at any other time in our history. In some cases there may be legal remedies which may be resorted to. In other cases it may well be that pressure groups may be necessary to bring influence to bear on the decision-making process.

The following are measures which can be used to assist in protecting various elements of the built environment:

- **Listed buildings** – legislation exists to protect buildings if they are seen to be of historical or architectural interest. Various grades are used which limits the amount and type of work that may be carried out on a particular building.
- **Conservation areas** seek to maintain the overall visual quality and character of the area. New buildings must fit in with the historical style surrounding it.

Table 1.3 Summary of features of town and city structures

Period	Features
Medieval	Rural society
	Market centres
	Local administration centres
	Market squares
	Ecclesiastical buildings
	Unplanned development
	Development around a well or cross-roads
	Narrow streets
Renaissance	Reduced control by the church
	Increased control by merchants
	Formal central square
	Geometric layout
	Long narrow squares
	Space around and between buildings
Georgian	Squares with town houses
	Opera houses and theatres
	Churches
	Development of spa towns
	Constraint on design and density of housing
Victorian	Villas and town houses
	Solid public buildings – town halls, libraries, museums, art galleries
	Industrial buildings
	Railway stations
	Back to back properties
	Garden City movement
	Absence of planning controls
	Industrial, commercial and residential areas
	Improved drainage systems
Modern	Muddled and congested central areas
	Commercial businesses in the centre
	Preservation of good buildings and open spaces
	Regulation of suburban development
	Broad highways and park land
	Control of character and design of buildings
	Indiscriminate development
	Guided industrial development
	Birth of New Towns
	Planning for motor traffic
	Neighbourhood shopping centres
	Zones of control

- **City grants** are available for specific instances to aid conservation of buildings or an area.
- **Town and Country Planning Act** seeks to control development or alteration of buildings and their use within the framework of a structure plan for an area, or a local plan for a more discrete area of the community.
- **Tree preservation orders** – planning authorities have the powers to issue tree preservation orders in order to retain the features of an environment.
- **National parks** seek to preserve the beauty of the countryside.
- **Green belts** prevent urban sprawl in areas of development pressure.
- **Environmental impact assessment** to be carried out on the granting of planning permission on larger industrial toxic developments.
- **Community architecture** aims to consult residents about their needs and keep them involved in the development of a project in their locality.

- **Pedestrianisation** seeks to either exclude or control the amount of traffic entering areas essentially designed as pedestrian precincts.
- **Housing associations** sought to acquire and manage properties in housing action areas.

Self-assessment tasks

The following questions relate to the nearest tov live:

1. Identify and describe the factors which influer location of the town.
2. Produce a line diagram of the town and identify on it the buildings and areas prior to 1850 and those post-1950.
3. Draw a sketch plan to show all the major transportation routes within your town and to other areas in your region.
4. On a line diagram of the town identify zones of use prior to 1850 and post-1950.
5. Prepare a five-minute talk on the importance of Town and Country Planning legislation.
6. Produce a collage to show the essential differences between a Victorian industrial city and a 'Garden City'.

...pment of construction materials
...ds

...rary construction

Basic needs of the human are food, clothing and shelter from the weather. Initially caves were used to satisfy the need for shelter, but these were not always convenient or available, and so the use of other materials was considered and attempted.

Those materials which were available and ready to hand were chosen. The principal ones were:

- timber
- clay
- mud
- grass
- reeds
- stone

These materials were naturally occurring and more widespread than caves. With the assistance of crude tools it became possible to fashion the basic materials into elements which were able to form a structure to satisfy the basic need for shelter.

As time progressed man explored and developed new and better systems of building construction often related to the advance of knowledge or techniques in using the same materials in different ways or in combination with other materials. Thus historians can demonstrate how some of the materials familiar to us today have been subject to change over time.

Stone

The third millennium B.C. saw stone being used to form step pyramids. Sedimentary sandstone or limestone were commonly used, sandstone being capable of spanning larger distances. In other locations igneous stone, for example granite, or metamorphic stone such as marble or slate were used.

Natural stones are strong in compression. Limestone and marble have low thermal expansion, but sandstone, granite and slate require allowances to be made for thermal movement. Other characteristics of these stones were discovered through practice such as:

- durability – resistance to water, chemicals, staining, frost
- workability – degree to which hand or machine is used
- permeability – fissures, absorption, grain

Examples of the use of stone can be found in most periods of history in:

- pyramids
- castles (Fig. 1.4)
- churches
- palaces
- bridges (Fig. 1.35)
- monasteries
- residences (Fig. 1.36)

The types of stone previously mentioned are still used today. What has changed is our understanding of their particular properties, the method in which they are obtained and worked, and the use of this knowledge to select stones for the most appropriate use and location.

A contemporary development is the manufacture of artificial stone. This is formed from crushed stones often in white or coloured cements which can be formed to resemble various natural stones in differing profiles. Where mass production is required casting in moulds is carried out. Strengthening may also be achieved by adding reinforcement.

Bricks

Bricks have been used for at least 8000 years. Originally they were formed by pressing mud into suitably sized lumps and leaving to dry naturally using air circulation, or the additional heat from the sun. Where clay was used it was prepared by

Figure 1.35 Masonry arch bridge, Bakewell, Derbyshire

Figure 1.36 Stone-walled residences at Bakewell, Derbyshire

Figure 1.37 A former hotel building in load-bearing brickwork, Norwich

adding water and treading in chopped straw or dung to prevent movement through twisting or warping.

Gradually wooden moulds were used to make the bricks and to maintain some form of standardisation. Little thought was given to the aesthetics of the brick. Eventually bricks were fired in various types of kilns which assisted in improving their strength and durability. Recognition that the type of clay affected a brick's properties and aesthetics led to greater care in the choice of clay.

In Britain the Saxons used Roman bricks and certainly up to 1325 bricks were imported from the low countries. A few buildings exist in East Anglia where local bricks were used. In the fourteenth and fifteenth centuries bricks were used for walls, pavings, fireplaces and flues.

After the Great Fire of London in 1666 brick was universally employed in the buildings of the capital. Mechanisation of brick production did not really develop until the middle of the nineteenth century when 20 000 to 40 000 bricks could be fired together.

In the last hundred years clay bricks were predominant in the United Kingdom, however the emergence of other brick types and concrete blocks has provided in some cases acceptable alternatives. Many different types of brick are produced to a standard metric brick size of 215 mm × 102.5 mm × 65 mm. The three basic types are:

- **common bricks** – a cheap brick not designed for high strength or finished appearance
- **facing bricks** – attractive in appearance and in some cases weather resistant (Fig. 1.37)
- **engineering bricks** – designed for strength and durability, usually of a high density

Concrete blocks

Concrete blocks are either dense or lightweight, lightweight blocks being one-third the weight of dense blocks. They are widely used for inner leaves of cavity walls, internal walls and lightweight partitions. Thermal conductivity varies with the density of the block. Insulation is provided by voids in blocks which can be filled with polystyrene to improve the insulation qualities. Fire resistance may be up to two hours.

The use of blockwork has:

- increased productivity on site
- the possibility of carrying structural loads
- removed the need for plastering or rendering in some cases.

Roof coverings

Thatch is mentioned in manuscripts as early as A.D. 700 where its inflammability was noted. As thatch has a relatively short life little is known of ancient methods of thatching. Several fires in the Middle Ages led to thatch becoming a forbidden material and therefore it began to fall out of favour except for where the availability of the raw material and the climate supported its use being maintained (Fig. 1.38).

Clay roof tiles were used by the Greeks and Romans, the Romans using them in Britain and elsewhere in their empire. It then appears that in the Saxon period thatch was the predominant roof covering which found itself being displaced by tiles towards the end of the thirteenth century. There were numerous irregularities in size and quality until standardisation took place in 1725.

The Romans used flat and round tiles, but the commonest tile in Britain was the plain tile. Early fixing methods consisted of oak pegs driven through holes in the tile and hooked over laths. In the nineteenth century nibs on tiles came into general use and have continued to the present day. Tiles are made in a range of colours and special shapes to suit particular situations e.g. hips, valleys, ridges of roofs (Fig. 1.39).

The practice of **tile hanging** over the underlying structure first appeared in the seventeenth century. In the Georgian period tiles were made which imitated bricks, some of which have survived to the present day. Tile hanging is still practised today as a weatherproofing method to common brick or blockwork, or on timber framed housing.

Stone slates or tiles were cut and used, though their weight probably limited extensive use (Fig. 1.40). **Concrete tiles** first appeared around 1840 in Europe but suffered from high porosity and deterioration of the colour. These present no problems in the modern tile, which is a viable alternative to clay tiles.

Shingles are thin slices of wood used in a similar manner to tiles to cover a roof. Early shingles were of oak, being used at one time on Salisbury Cathedral. Like thatch they posed a fire risk and so their use declined at a similar time to thatch. Shingle roofs can last up to one hundred years though their use is rare today. Where shingles are used then Canadian cedar is chosen.

Slate was used as a building stone in earlier periods, which led to the discovery that it could be split along parallel planes

Figure 1.38 Timber-framed building, dating from the fourteenth century, with a thatched roof

Figure 1.39 Clay pantile roof

quite easily. This enabled thin 'slices' of slate to be created which we call roofing slate. Slate's main qualities of being close textured, non-porous and resistant to frost soon led to it being used as the cladding medium on roofs.

Figure 1.40 Stone slab roof. Note the different sizes

Initially the Romans used slate, which declined in use for a time mainly due to transportation difficulties and its own dead weight. It did continue to be used in localities where it was naturally occurring. Colours of slate from these areas are:

- The Lake District – grey and grey–green
- Wales – dark and light blues, light and dark green, dark grey
- Cornwall – grey and grey–green

Welsh slate was used in Chester and eventually spread further afield, being transported initially by sea, and then as they developed by canal and railway. Slate continues to be used as a roofing material. It is also used in a similar manner to clay tile hanging as a vertical protection to building surfaces (Fig. 1.41).

Modern uses are for paving, cladding to framed buildings, roofing and damp-proof courses. It is also used in prestigious buildings carved in low relief as memorial and other types of plaques.

Glass has increased in popularity as a roof material in modern buildings, particularly with glazed atrium shopping malls (Fig. 1.42).

Figure 1.41 Modern slate roof covering

Figure 1.42 Roof glazing at the Castle Mall shopping centre, Norwich

Timber

Timber is one of the oldest materials used for building purposes. Initially in its most basic form branches were used to form what we now call flat, lean-to or pitched roofs. The branches were interwoven into an **'A' frame** and acted as the structural members and were then in-filled with either straw, reed, or heather.

Examples of this are in the Norman period where a **cruck frame** for a small hall was common. This used a pair of trees split down the middle supporting a ridge piece. They were connected at a suitable height by a tie beam. Eventually the extended tie beam supported a wall plate which in turn supported rafters and a roof covering.

As larger spans became a requirement larger sizes of timber were used and were roughly fashioned into square or rectangular shapes. Timber buildings were commonplace adjacent to wooded areas in the country. Oak was a predominant material, being very durable with examples surviving to the present day.

The development of hand tools enabled the medieval builders to fashion and shape the wood and be able to produce carpentry joints which are still used today. The use of jointing allowed longer and sometimes stronger lengths of timber to be used which increased the spanning capability of timber even further. Timber was also used to form the internal and external walls of a structure and so the early timber frame building was born with the stud partition (Fig. 1.43).

The nineteenth century saw steel being introduced, and designers, realising the strength characteristics of it, used it sandwiched and bolted in between two timbers, which increased the beam's resistance to bending. This enabled timber floors to span larger unsupported distances than previously.

In the twentieth century there was greater understanding of timber and adhesives technology which led to the development of laminated timber. This could enable timber to compete on similar terms to materials such as concrete and steel in framed buildings.

Figure 1.43 Timber-framed building undergoing renovation in Suffolk. The vertical studs and internal lath and plaster wall finish are clearly visible

Cement, plaster and mortar

Cements are hydraulic materials in that they depend upon water rather than air for strength development. When water is added to cement a chemical reaction takes place and continues while water is present. Hydraulic cement stiffens first and gains strength later.

In 1796 James Parker found that he could make a hydraulic cement by calcining nodules of argillaceous limestone from the London clay cliffs. This cement was called **Roman cement** and was used till the middle of the nineteenth century. In 1811 a hydraulic cement was made by calcining a mixture of clay and limestone. Improvements to this were made until the first **Portland cement** was made in 1845. It is of course an important ingredient of **cement mortar** and **concrete**.

Figure 1.44 Seventeenth-century pargetting at Clare, Suffolk

The finest **plaster** is produced by burning gypsum which when subjected to heat produces **plaster of Paris**. Henry III introduced it to England in the thirteenth century from France. Soon gypsum was being quarried in England around Purbeck and in Yorkshire. Plaster was used for both floors and walls where it was spread on to reeds on timber laths. Plaster was also used as an external wall covering initially to provide additional protection but later for aesthetic reasons.

Pargetting was a term used to describe ornamental designs in plaster relief or by extension incised. It is associated with timber frame buildings of the seventeenth century. Its use declined as timber frame buildings lost favour (Fig. 1.44).

Mortar is a loose term which describes a material used in brickwork and masonry, and consists of a binding material and a filler. The earliest use of mortar was to cement together the sun-dried bricks of early structures in Egypt. **Lime sand mortars** were used in medieval building. The mortar had relatively little strength and hardened slowly.

Lime mortar was used up until the start of the twentieth century when the much improved Portland cement replaced the lime as the principal binder. Various types of mortar have been developed which provide varying degrees of strength and durability. Recently **lightweight mortars** have been introduced which have a much lower heat transmission than the other mortars.

Lead

Lead was the only metal in great demand prior to the Industrial Revolution as a roofing material, and for use in containing glass in windows. It is a soft and pliable substance with no weight-bearing ability at all. England possessed large lead reserves, so much so that the bulk of the world's supply was from England.

Until the end of the seventeenth century lead was sand cast in heavy sheets, where it then became possible to roll the lead and produce uniform thicknesses. Lead was principally used for roofs, being used on many religious buildings. From the

Tudor period lead was used for internal plumbing and more frequently for gutters, down-pipes and rain-water heads (see Fig. 1.17), lead down-pipes being used in the Tower of London in 1240. Eventually other materials replaced lead for water supply, present day substitutes being copper and PVC.

Iron

There were three distinct types of iron:

- **cast iron**
- **wrought iron**
- **steel**

Abraham Darby revolutionised the manufacture of **cast iron**. The large scale production of **wrought iron** was made possible by Henry Cort, and Bessemer, Siemens and Thomas invented methods of producing **steel** cheaply and in large quantities. Steel was harder wearing and more tensile than either wrought or cast iron so it assumed greater importance in the eighteenth and nineteenth centuries.

The structural use of cast iron in buildings began with attempts to increase the fire resistance of mill buildings. It was used for columns, beams and window frames, and stone-flagged floors were carried on brick arches between iron beams.

While the fashion for cast iron was still at its height in the 1850s there was a considerable resurgence in the use of wrought iron. This in part was due to the fact that the manufacturing process had been improved to produce a purer iron. This allowed the rolling mills to produce small 'I' beam sections for use in floors.

Prior to 1850 steel production had been limited to the production of objects of a moderate size such as tools and springs, as the process was expensive and the output small. Bessemer's process made production of steel in large quantities possible and so the production of beams of a substantial section was possible. Many steel mills moved to the production of beams and a large range of non-standard sizes was available.

In 1904 the forerunner of the British Standards Institute sought to restrict the range of sizes available and also the quality of the steel from which the beams were made. In 1877 the use of steel in bridges was approved and so the Forth bridge used 50 000 tons of steel. The first wholly steel framed building was probably the Rand-McNally building in Chicago in 1890.

Concrete

Lime concrete was used by the Romans, being mixed sometimes by pouring grout over broken stone or brick, and sometimes with lime, sand, broken stone or bricks and water. This use continued at intervals throughout the Middle Ages.

From 1850 onwards concrete was made with the newly developed Portland cement, which was much preferred to the other artificial cements. Mixing was mainly by hand except on the very large jobs where steam driven mixers were used.

It was recognised that concrete could be used as an alternative to masonry and that it was strong in compression and weak in tension. With masonry attempts had been made to overcome this tensile weakness by using iron rods embedded in the joints, which had proved unsuccessful. To put iron bars in concrete was far easier and so the birth of reinforced concrete occurred. In addition to improving the tensile qualities of concrete, the use of Portland cement increased its weatherproofing properties.

In 1888 E. L. Ransome embedded iron beams in concrete to make the first reinforced concrete beams. By the end of the century he was developing systems based on unit construction. As understanding grew concrete quality improved, this being a result of better compaction and the careful control of the proportions of water and cement in the mix.

Lightweight concretes were introduced after the second world war and extensively used in multi-storey construction. Pre-stressed concrete was a further innovation and of major structural importance. Pre-stressing had been carried out as early as 1811 but had not been too successful as steel of a higher strength than normal mild steel was not available.

Freyssinet is credited with its early development which coincided with a better understanding of normal reinforced concrete. He concentrated on developing and improving the anchorages for the stressed reinforcement in order to improve the efficiency of the beams. The use of pre-stressed concrete in buildings up to the 1950s centred mainly on use as a replacement for normal reinforced concrete.

Self-assessment tasks

1. Describe the changes that have occurred in the development of materials from early times to the present day.
2. Take a walk around where you live, noting down the materials that are used to form the buildings, and the type of building they are used for.
3. Describe the influence that improved transportation had on the use of materials.
4. Describe the advancements that materials made in the late nineteenth century, and suggest reasons why they occurred.
5. Explain how the refinement of concrete led to changes in the construction of buildings.

The use of early and contemporary materials in the construction of building elements

It is possible, thanks to surviving buildings, remains or historical records to be able to picture how early buildings were constructed. The method of construction and the materials used in early times did very much rely on the locality in which the building was situated and who the building was for. This led to the development of different approaches to construction, and even different practices with the same materials. This isolation gradually disappeared as travel about and between countries increased and ideas and materials from one area became transferred to another.

Observations of buildings built throughout the various periods will reveal that they have developed enclosing structures using what we now call building elements. For the purposes of this unit those elements are:

- foundations
- walls
- floors
- frames
- roofs
- drains

Foundations

Some buildings were built directly on to the earth at ground level, and even in the 1920s foundations on domestic properties were as little as 600 mm below the ground level. Medieval builders recognised the need to place buildings on a firm base and therefore formed foundations in prepared trenches into which a well compacted layer of stone, chalk or flint was laid. These materials were often bound together by strong lime mortar. The building structure of walls or columns was then built directly from this prepared base.

Timbers such as oak, elm and beech were used in marshy wet areas. The timbers were driven down into the ground until a firmer strata was reached. The use of hardwoods ensured that they would last longer than softwoods. A preservative method used to extend the life of the timbers was to char or scorch the wood.

Early bridge foundations used timber piles driven into the river bed in clusters which were held together by planks of elm or oak. Planks also rested on top of the piles and were used to support the masonry of the bridge construction.

As bricks became the predominant walling material the tradesmen recognised that bonding could be used to effect by spreading the load of the wall over the ground. This practice was known as doing the 'footings' (Fig. 1.45). It was used until concrete with Portland cement became commonplace, when the strip foundation took over.

Concrete replaced timber as the main foundation material. As steel became available and was compatible with concrete it enabled the load-bearing characteristics of foundations to be improved, especially in weaker ground. Where buildings were to be built on weak soil, or in water, reinforced concrete piles were used.

Walls

The main functions of a wall are to:

- provide security
- protect from the weather

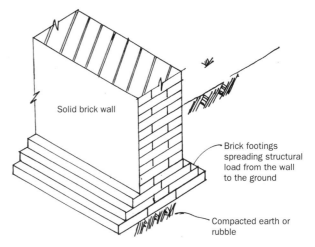

Figure 1.45 Brick footings

- carry loads of various kinds
- divide up a space

In primitive dwellings wall and roof construction were virtually the same element. As has been previously described the cruck frame was made from stout timbers as rafters or studs on to which branches were intertwined to support a further water resisting layer of mud, straw and reeds. Later timber-framed walls were used, only to be replaced by stone or brick.

Stone had always been in great demand for the most prestigious buildings, and if not locally available was transported great distances from its place of origin to the point of use. This was particularly so for cathedrals and the palaces. The larger stones were used at the base of the wall as their physical size prevented them from being used elsewhere. Their size also helped to take and spread the load from the walls to the ground, thus acting as a form of foundation. Inner and outer walls were formed and were interlinked by deep bonding stones, the void being filled with poorer quality stone.

Flint, cobbles and pebbles were used as an alternative method of construction but posed problems due to their irregularity of shape (Fig. 1.46). They were often tied together by 'lacing courses' of brick or stone. They were heavily reliant on great quantities of mortar and backing, or the use of other supporting materials to 'reinforce' the stability of the wall (Fig. 1.47).

Earth in the form of clay, cob or wychert (a type of earth and chalk mix) was used for walls in certain areas of the country, some surviving to the present day. Cob buildings are a feature of Devon (Fig. 1.48). The walls used a mixture of straw and manure which when dry were coated with stiff clay or lime plastering. The walls were of sufficient strength to be able to support floors and roofs.

The use of brick in walls spread rapidly from the late eighteenth century. The regularity of the shape of a brick did not prevent the builders from using it imaginatively to produce many variations and textures to the facades of brick buildings. This was achieved by using:

- different bonds (Fig. 1.49)
- various types of bricks
- varying features such as arches over doors and windows
- a mixture of jointing techniques

It was not until the twentieth century that cavity walling and the use of other materials such as blockwork for the inner leaf occurred.

Contemporary framed buildings use various methods and materials to perform the enclosing function. Figure 1.50 shows brick cladding to a multi-storey car park. Figure 1.51 shows curtain walling to an office block and Fig. 1.52 shows infill panels to an office building in London.

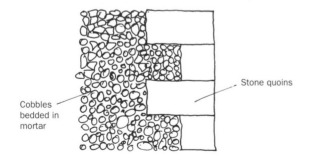

Figure 1.46 Cobbled wall

Cobbles bedded in mortar

Stone quoins

Figure 1.47 Wall built from flint with brick lacing courses and quoins

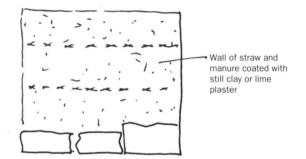

Wall of straw and manure coated with still clay or lime plaster

Figure 1.48 Cob construction

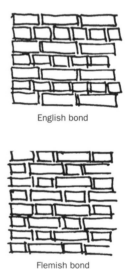

English bond

Flemish bond

Figure 1.49 Brick bonds

Frames

The use of the cruck frame (Fig. 1.53) was not widespread throughout England, being limited to the West, Midlands and the North. The most popular was **box frame** construction (Fig. 1.54). Here the structural members also formed part of the wall, with the wall fabric forming the enclosing element to keep out the weather, and maintain some privacy and security. The main timbers of the frame, called studs, could be exposed or concealed.

Figure 1.50 Decorative brick cladding to a multi-storey car park, Chesterfield

Figure 1.52 Infill panels to an office building, London

Figure 1.51 Curtain walling to an office block, Norwich

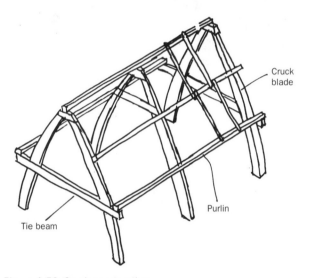

Figure 1.53 Cruck construction

Timber-frame building is used today for domestic properties (Unit 3). Instead of brick infill, bricks are used as an outer leaf to the inner timber frame. The frame is clad with sheeting of either plasterboard and skim, or some other manufactured sheet such as plywood, or tongued and grooved boarding. An alternative to the brick outer skin is the use of tile hanging as the outer waterproof layer.

Roofs

In cruck framed construction the roof was part of the total structural system. All other structures after cruck buildings used a roof construction which was a separate part of the structural elements of a building.

Roof constructions varied according to the function of the roof and the materials used to form the roof surface. While one of the purposes of a roof is to prevent weather penetrating into the interior, other uses are for the roof space to be used usefully as a room. This has led to roofs becoming known as gabled, hip, mansard (Fig. 1.57) or gablet, the construction of the roof determining how much useful space will be provided (Fig. 1.58).

Vertical studs formed the main frame which was infilled with other studs (Fig. 1.43). Not all studs carried structural loads. Horizontal members such as the cill, wall plate or tie-beam completed the frame. Between the timber members bracing was placed to give some stability to the frame.

In the Middle Ages it was usual practice to infill the frame with wattle and daub panels, which were subsequently replaced by brickwork (Fig. 1.55). Additional height to a timber-frame building caused a problem with floor stability. This was solved by using 'jetty' construction where each successive floor projected beyond the frame of the wall beneath it (Fig. 1.56). The use of jettying allowed the floor area on the upper floors to be increased. With the passage of time the timber frame beneath jetties rotted and was replaced with brick or stone.

Figure 1.54 Box frame construction

Figure 1.56 Jetty construction

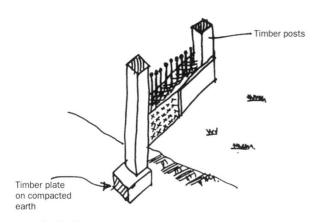

Figure 1.55 Wattle and daub construction

Figure 1.57 Mansard roof with interlocking tiles. Semi-detached property, Chesterfield

In addition construction of the roof structure centred on whether the rafters would provide their own stability by the way the roof was constructed, or whether the roof had rafters which transferred their load to roof trusses. Roof construction was a highly developed form, particularly in churches and other large buildings where the use of trusses allowed long clear unsupported spans between supports (Fig. 1.59).

The aesthetics of roofs were determined by the roofing material. Flags and slates, and thatch and tiles create their own particular character for a roof, particularly when used with dormer or eyebrow windows or low-swept eaves.

Modern roof construction uses a mixture of all the methods developed over the years. As steel became available so did trusses fabricated from steel sections which supported purlins to which were fixed the rafters supporting the roof finish. Roof finishes still continue to be made from clay, slate, concrete (Fig. 1.60), timber and metals (lead and copper) with the addition of bitumen felt, and bitumen slates and various plastic profiled sheets.

Floors

The construction of floors varied according to the type of building. Domestic medieval buildings would often be single storey, so the ground floor could be composed of consolidated earth. Baked clay tiles were made between the twelfth and sixteenth centuries from red clay mixed with sand. Often they had a glazed surface but as the firing temperature was so low they were not durable and perished. Regional differences also occurred so it was not unusual to find stone flags in one area and slate in another.

Buildings with more than one storey had floors of timber

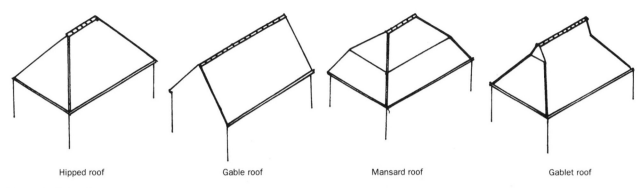

Figure 1.58 Roof types

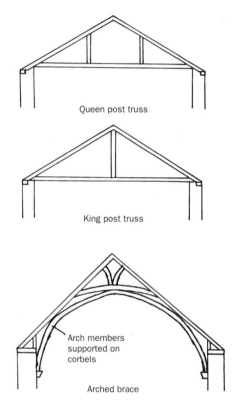

Figure 1.59 Truss construction

Figure 1.60 Concrete interlocking tiles being laid on a bungalow roof

boards supported on joists spanning between walls or other forms of support such as larger section size timber beams. This form of construction was used with timber-frame construction, masonry and brickwork. Stone and brick vaulting was also used, more so over cellars or crypts and vaults.

When rolled steel joists became available it was possible, because of advances with other materials such as concrete, to use a composite construction of the two materials to produce a floor which could span longer unsupported distances and higher loads. A much valued property of fire resistance was now possible.

Drains

The disposal of sewage from properties of the well-off in medieval times was to cesspits which normally served the one property. In most cases though sanitation was provided by **privies**. Newer properties might have had one per household but it was not uncommon to find up to fourteen families sharing one privy.

The soil holes were usually open and became offensive in use. They often overflowed causing misery and suffering for the inhabitants. During rainy weather the privies were often emptied into gutters in the streets. In some areas privies were emptied at night times to local dung-hills.

In general, lack of proper sanitation led to repeated epidemics and widespread disease. Overcrowding and the contamination of water supplies by cesspits made conditions worse. By the end of the eighteenth century the **water closet** in a basic form was being introduced. Its introduction led people to believe that the sanitation problems had ended. In fact this was not the case, odours being directly introduced into the house. In 1782 the first patent for a **trap** was taken out.

Advances in sanitary engineering continued until at the end of the nineteenth century satisfactory methods of connecting to a sewer were brought into use. This was by using the **two-pipe** system of drainage (Fig. 1.61) which allowed the contents to be discharged to a drain or sewer. These sewers were often inadequate, being undersized and laid at shallow gradients with blockages being frequent. They discharged untreated effluent into ditches, rivers and streams which caused further pollution problems.

In 1848 the introduction of the **Public Health Act** signalled the start of modern sanitary law. The **Town Improvement Act** of 1847 authorised the appointment of a person to act as a surveyor of paving, drainage, etc. This was subsequently embodied in the **Public Health Act** of 1875.

In 1855 the **Metropolitan Board of Works** was created which assumed responsibility for London's sewers. A network of brick sewers (Fig. 1.62) was subsequently developed which took raw sewage to suitable discharge points or treatment

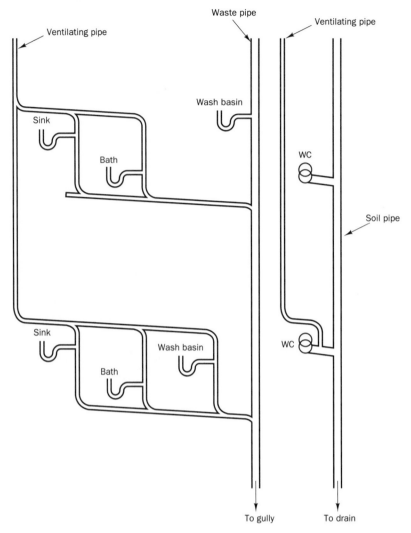

Figure 1.61 Two-pipe drainage system

plants. Initially this building and sanitary legislation had been confined to London but in the 1850s it was appreciated that the extension of the legislation to the whole country was long overdue. Thus many large towns and cities developed their own sewerage systems.

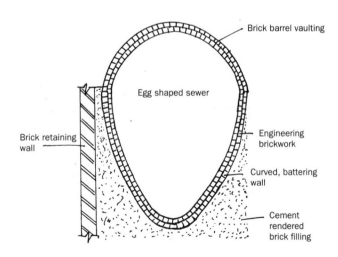

Figure 1.62 Typical brick sewer

Self-assessment tasks

1. Explain the following terms using diagrams where appropriate:
   footings, jetty, cruck frame, King Post Truss.
2. Describe the differences between a wall from the Elizabethan period and one from the twentieth century.
3. Describe with the aid of sketches how timber has been used in domestic construction from the Saxon period.
4. Explain how the disposal of sewage contributed to unhealthy living conditions in the late nineteenth century.
5. Consider the property you live in. Note down the different materials that are used to form the various building elements, and whether they existed prior to the twentieth century.

The influence of early construction methods on contemporary construction methods

Advances and improvements in the methods of construction are a result of:

- adopting good practices from the past
- development of new materials

- increased understanding of the behaviour of materials and building structures
- introduction of minimum standards, often through legislation
- improved manufacturing techniques and quality control

This section seeks to summarise those points which should be considered when reviewing the thinking or influences on the way buildings are constructed. They are presented by addressing the major elements of a building. Aids to expanding these points can be found elsewhere in this volume, and also in *GNVQ Construction and the Built Environment – Advanced* published by Addison Wesley Longman. Contemporary methods in this context have taken a date line of post 1900.

Foundations

Early methods:

- foundations built directly on to the earth
- foundations built on to compacted fill
- use of large stones as base course of walls
- introduction of concrete
- use of steel reinforcement
- use of brick footings as spread foundations
- foundations taken below ground to minimise climatic influences
- hand mixing of concrete
- timber pile foundations

Contemporary methods:

- scientific concrete mix design
- predictable behaviour for different foundation types
- mechanical mixing and placing of concrete
- minimum standards through Building Regulations and British Standards
- ground compaction techniques on weak soil
- projection of foundation beyond the wall's width
- concrete pile foundations
- use of sulphate-resisting cement

Walls

Early methods:

- materials of stone and brick
- wattle and daub construction
- timber-frame construction
- use of bricks
- lime mortar for bedding bricks
- solid construction
- tile hanging
- relationship of mass to stability
- flying shores and buttresses
- hand manufacture of bricks
- labour intensive construction methods

Contemporary methods:

- cavity wall construction
- use of lightweight blocks
- improved thermal insulation properties
- reduction of the cold bridge effect

- developments in supporting openings e.g. lintels
- lateral restraint to walls from floors and roofs
- stiffening walls with piers and buttresses
- British Standards for brick manufacture, cement
- use of slenderness ratios to aid wall stability
- introduction and use of fire barriers
- use of damp-proof courses at various positions in a wall
- Building Regulation statutory requirements
- mass production of bricks and blocks
- labour intensive construction methods

Frames

Early methods:

- cruck construction
- wattle and daub
- timber frame
- pargetting and stucco
- stability governed mainly by one timber and its size
- joints all constructed by hand on-site
- use of flitch beams
- jetty construction

Contemporary methods:

- timber-frame construction
- use of moisture and vapour barriers
- tile hanging
- laminated timber
- use of steel frames in various formats, e.g. portal frame
- use of reinforced concrete frames
- timber preservation techniques
- welded and bolted methods of construction
- pre-fabrication off-site
- use of different cladding materials, e.g. GRP, concrete, steel
- plasterboard and other dry lining materials as an internal finish

Roofs

Early methods:

- 'A' frames
- rafter and purlin construction
- truss construction, e.g. King and Queen post trusses
- coverings of slate, concrete and clay tiles, lead sheet
- traditional jointing methods, e.g. mortise and tenon, scarf joints, bolted
- timber selection by sight inspection
- triangulated construction for strength and stability
- laborious hand timber conversion techniques
- site-based construction and assembly methods

Contemporary methods:

- rafter and purlin
- truss, rafter and purlin
- trussed rafter
- machine stress graded timber
- mechanical jointing methods, e.g. timber connectors
- laminated timber construction
- off-site pre-fabrication techniques
- mechanical handling on-site

- wind bracing and lateral restraint
- introduction of steel roof trusses
- increased accuracy with timber conversion and sizes
- influence of Building Regulations, e.g. stability, sizes, grade of timber
- use of fire breaks
- use of a wide variety of roof finishes appropriate to the roof form
- increased accuracy in setting out and construction

Floors

Early methods:

- solid ground floors of earth, stone or tile
- no formal damp-proof course
- no ventilation under suspended floors
- no oversite concrete
- use of unseasoned timbers
- timber selection by sight
- jointing by traditional hand methods
- size of joists determined by custom and practice
- minimal legislative requirements to be satisfied

Contemporary methods:

- suspended ground floor construction improved by compliance with Building Regulations and the Approved Documents
- only seasoned timbers used
- selection of timbers for structural strength by mechanical stress grading
- joints can be formed by using mild steel joist hangers
- use of double floor construction to increase spans but by using steel beams
- sound- and fire-resisting construction
- use of manufactured boards as timber substitute for deckings
- introduction of concrete beam and blocks for suspended floors
- use of reinforced concrete for suspended floor slabs
- use of damp-proof membranes below ground floors

Drains

Early methods:

- disposal to cesspits and ditches
- no sanitary equipment
- use of shared facilities
- open earth closets
- lead pipework

Contemporary methods:

- legislation introduced to control sewage disposal
- sanitary equipment developed, particularly the water closet (W.C.)
- water traps introduced to prevent odours etc. reaching the building interior
- design of above and below ground drainage systems
- discharge of sewage more controlled
- sewage treatment plants introduced
- emphasis on separate drainage systems
- development of good quality drainage materials and joints

Self-assessment tasks

1. Explain, using examples, how the Building Regulations have influenced the improvement in standard of building elements.
2. Describe using two examples, how the increased knowledge of materials has led to improvements in their use.
3. Describe the factors that have led to standardisation of building components.
4. Sketch a brick external wall of the 1850s and one of the 1990s. Describe their essential differences and suggest reasons for the changes.
5. Describe how early timber-frame construction has influenced that of the twentieth century.

Early and modern civil engineering structures

Bridges

Bridges have been used since very early times when the need to cross water arose. Initially the alternative was a ford, but gradually the use of a **packhorse bridge** became commonplace. Materials used were timber or stone. In some areas slabs of stone were laid across between supports forming perhaps the earliest **beam bridge**, but by far the most popular was the **arch bridge** built of masonry (see Fig. 1.35). Later developments saw the introduction of steel and concrete used in **suspension** (Fig. 1.63), **cable stayed** and **cantilevered bridges**.

Arch bridges

Arch bridges support their imposed load by carrying the weight outwards along two curving paths to the abutments where the forces are resisted by the ground and assist in maintaining the stability of the bridge (Fig. 1.64). Stone and brickwork were used to form elliptical, segmental and semi-circular arched bridges. As spans increased the height of the bridge would also increase. In order to limit this multi-span bridges between supports were constructed (Fig. 1.65).

As methods of transport changed and canals and later railways became common the need for bridges increased. The building of arch bridges in brick or stone was lengthy and time consuming. Materials had to be transported to the sites often in inhospitable locations. Large timber supports called **timber centres** had to be constructed first. These were temporary constructions which supported the bridge under

Figure 1.63 Simple braced-beam suspension bridge, St Olave's, Norfolk

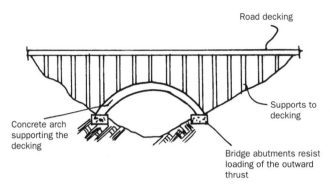

Figure 1.64 Arch bridge

Figure 1.65 Reinforced concrete multi-span bridge

construction until the mortar had fully set. The centres were then **eased** to gradually allow the structure to take up its full load and then **struck** and removed completely. Skewed brickwork was often used to increase the strength of the bridge.

Other materials have been used to create arch bridges, one of the most famous being the Ironbridge in Shropshire which was the first iron bridge in the country. The world's most famous bridge is the Sydney Harbour Bridge which was a steel through arch bridge built in 1932. Also in Sydney is the longest concrete arch bridge built in 1964 spanning 305 metres.

Beam bridges

Various materials have been used to form beam bridges but they all have to overcome the same problem, that is, how can the deflection of the beam supporting the platform be minimised? Clearly materials which are weak in tension are unsuitable for this unless their weakness can be overcome as is the case with concrete.

Steel is a versatile material and has been used in many innovative ways to provide sufficient stability to varied types of beam bridge construction. The span and loading of the bridge tend to be major influences on the bridge design and how many intermediate supports may be needed. As the name suggests beam bridges use beams as the main structural member but there also needs to be a deck to suppport the road or railway.

Common forms which have been used in beam construction are:

- simple 'I' beam sections
- hollow beams, or tubes
- braced girders
- plain reinforced concrete
- pre-tensioned or post-tensioned concrete (Fig. 1.66)
- box-girder deck

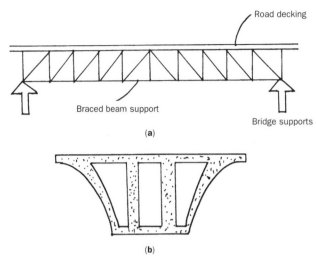

(a)

(b)

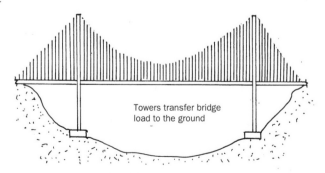

(c)

Figure 1.66 Beam bridge: (a) steel beam support; (b) pre-cast concrete beam segment which is post-tensioned to other segments to form the bridge; (c) pre-tensioned concrete beams supporting a road slab

Suspension bridges

The suspension bridge consists of a relatively shallow deck, which supports traffic movement, spanning between towers. The deck, being of a cross-section which cannot support itself, gains its support from steel hangers which are suspended from a continuous cable suspended from the top of the towers. These cables carry the weight of the decking to the top of the towers which transfer the load to the ground (Fig. 1.67).

Cables of modern bridges are made up from many thousands of steel wires which are bound tightly together. The steel which is used for the suspension cable and the hangers is able to resist the tensile stresses very effectively. This fact allows suspension bridges to be able to cover very long spans.

An early bridge built by Brunel was the Clifton Suspension

Towers transfer bridge
load to the ground

Figure 1.67 Suspension bridge

Bridge near Bristol in 1862. This spanned 702 feet. The deck was suspended from wrought iron chains previously used on a bridge at Hungerford.

A cross between the beam and suspension bridge is the **cable stayed bridge**. It was first used in 1956 in Sweden. The deck is supported from cables connected directly to the towers, unlike the suspension bridge which uses a suspension cable. A major advantage of this bridge is that it uses fewer piers than a beam bridge, and is more suitable than a suspension bridge over shorter spans.

The Humber Bridge, opened in 1981, is the world's longest single-span suspension bridge with a main span of 1410 metres. The bridge deck (roadway) hangs from huge suspension cables suspended from towers 155 metres high.

Other suspension bridges are:

- the Forth Road Bridge, opened in 1964
- the Severn Bridge (988 metres), carrying the M4 motorway, opened in 1966
- the Golden Gate Bridge, San Francisco, 1280 metres long, opened in 1937

Figure 1.68 Stone retaining wall, Derwent Dam, Derbyshire; completed 1916

Dams

Dams are civil engineering structures used to store water. They are normally associated with being built across a valley in order to retain the water trapped behind the dam wall (Fig. 1.68). Their early role was to store water to irrigate the land, but as a result of the industrial revolution dams were built to provide water for drinking and washing purposes, and for industrial purposes. Latterly they have been built to provide hydro-electric power and water supplies.

Dams are separated into two main types:

- **embankment dams**
- **concrete dams**

Prior to 1800 the early dams were all of the embankment type being built from earth or stone found near to the construction site. The oldest embankment dam is in Jordan dating from around 4000 B.C. It was built from earth with a stone facing to it.

Embankment dams are sub-divided into **earthfill** and **rockfill** dams, though the use of both materials was not uncommon in the same structure. Further types relate to the method used to create the waterproofing element. These were:

- central clay core
- sloping clay core
- upstream membrane of asphalt or concrete

The prevention of water seepage was achieved by using **puddled clay**. This was formed by using a suitable naturally occurring clay which after being made possibly more workable by adding water and sand was compacted to form the dam wall. Early dams built by this method have become unserviceable because of the high leakage rate and risk of collapse.

As the population of the world increased, bringing with it the ever rising demand for water, more dams were constructed which were larger and the walls went higher and higher.

- Grand Coulee Dam – a large gravity dam built in 1942 on the Columbia River in Washington. It is 1272 metres long at its crest and 108 metres high. It provides flood control measures and water for irrigation and hydro-electric power.

- **Aswan High Dam** – on the River Nile in Egypt. It was completed in 1970 and is 100 metres high. It provides water for irrigation, domestic and industrial use and hydro-electric power.
- **Grand Dixence Dam** – a gravity dam built on the River Dixence in Switzerland. It has a wall 284 metres high and 670 metres wide.

At the time there was a limit on the extent to which embankment wall stability could be assured which coincided with advances in concrete technology, and so construction of concrete dam walls became popular.

Concrete dams are divided according to the method they use to remain stable and are:

- Gravity dams – these are the simplest as they rely on their gravitational force to resist the pressures caused by the stored water upstream (Fig. 1.69).
- **Hollow gravity dams** – these used less concrete which reduced the cost of construction, but required careful design in order to maintain stability from overturning and sliding.
- **Buttress dams** – dam walls could be relatively thinner in width but in order to maintain the stability of the face concrete buttresses were used at intervals across the dam.
- **Arch dams** – these are constructed using a curved wall, the convex face being upstream, spanning between the valley sides. The wall is then able to resist the large thrusts caused by the volume of water stored. The face may be single or double curvature (Fig. 1.70).

The latter part of the nineteenth century and early part of the twentieth saw a reduction in the number of embankment dams in favour of concrete. That trend has now been reversed as technology will allow embankment

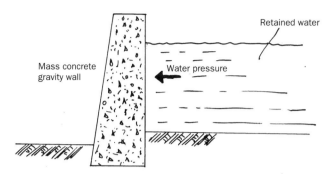

Figure 1.69 Gravity dam

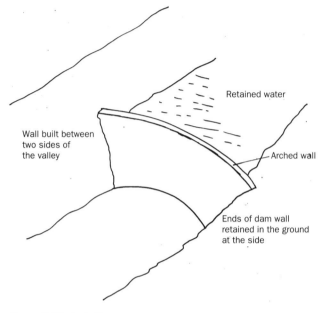

Figure 1.70 Arch Dam

dam walls up to 325 metres in height. This change in trend is attributed to:

- improved understanding of the behaviour of embankment dams and advances in soil mechanics
- the increased capacity of earth-moving machinery
- a reduction in the number of sites with bedrock suitable for concrete dams
- the increasing cost of labour

Even sites with strong bedrock near the surface have used embankment dams in preference to concrete. Embankment dams can tolerate poor ground conditions and their low cost of construction makes them a financially better option. Modern technology has also reduced the need to excavate to form **cut-off walls** in poor or permeable ground. Contemporary methods allow the formation of a **grout curtain** which reduces the permeability of the ground and allows some support to the dam wall by strengthening the foundation.

Self-assessment tasks

1. In your own location find two bridges made from brick, stone, timber or concrete. Sketch their shape in outline and classify them according to the bridge types.
2. Explain why embankment dams fell out of favour and then reappeared in the twentieth century.
3. Use your library to find out more about the Clifton Suspension Bridge and the Humber Bridge. Describe their construction and explain why the Humber Bridge could be built with a larger clear span.
4. Describe the construction of two beam bridges – one pre-1950 and the other post-1950.

1.3 Employment and career opportunities within construction and the built environment

Many people think that the built environment has nothing to do with them, yet all around them twenty-four hours a day, whether living, working, at leisure or sleeping they are within the built environment. So what is it that the built environment provides for these people?

Firstly, the buildings, which have been designed to provide accommodation which they will recognise as:

- houses, flats, bungalows and maisonettes
- shops, offices, hotels and care homes
- factories, lock-ups, warehouses
- sporting and leisure centres
- schools, libraries and churches
- surgeries, chemists and hospitals

Secondly, the transportation links they use such as:

- roads, motorways
- railways
- canals
- airports

Thirdly, the activities which maintain that environment:

- maintenance
- refurbishment
- renovation

Lastly, the integration of the built environment with the natural environment:

- rivers, lakes
- environmental protection
- legislative controls
- national parks, green belts

All of these are part of the very large world of the construction industry, and when looked at from this viewpoint there are a lot of people who are either directly or indirectly connected with that industry. The industry involves so many different activities and people that there is a career to suit almost everyone.

What follows introduces the relationship of the provision and delivery of those items noted previously, the personnel involved and their roles within their respective organisations, and the training and career routes that are available.

Activities and organisations in the built environment

Construction activity in the built environment relates to two main work divisions:

- those associated with new building work
- those related to maintenance, adaptation, restoration and refurbishment.

Table 1.4 Main work divisions

Type of work	Main work
New	A proposed building or one that has recently been constructed; roads, reservoirs, dams
Maintenance	Repairs to a building or its services to a standard which will allow them to continue to function
Adaptation	The alteration of an existing building so it may fulfil a different use
Restoration	Work carried out to bring an existing building back to its former condition
Refurbishment	Work carried out on a building which may involve adaptation and restoration in order to meet present day standards

Typical activities are shown in Table 1.5

Table 1.5 Typical construction activity

New work	Repair and maintenance
Houses	Doors
Shops	Windows
Offices	Roofs
Hospitals	Drains
Roads	Heating
Libraries	Electrical re-wire
Schools	Re-decoration
Churches	Lifts
Sports complex	Refrigeration plant
Theatres	Insulation
Sewers	Re-surfacing
Major bridges	

These activities involve work which is undertaken for clients who are either from the:

- public sector or
- private sector

The **public sector** consists of work normally financed from public funding and commissioned by:

- local authorities
- central government departments
- statutory authorities
- public undertaking

The **private sector** consists of work financed by private funds from individuals, or a company. These clients include:

- private individuals
- partnerships
- companies

Activities within these sectors are shown in Table 1.6

Table 1.6 Activities in the public and private sectors

Public	Private
Housing	Housing
Non-housing	Commercial
Repairs	Industrial
Maintenance	Repairs
Civil engineering	Maintenance

Six main work areas are able to serve the two sectors in order to meet their needs. These are shown in Table 1.7.

Table 1.7 Main work areas contributing to the built environment

Work area	Contribution
Building	Design, construction, maintenance and adaptation of buildings. Includes homes, leisure centres, hospitals, schools, office blocks, hospitals and industrial estates.
Civil engineering	Design and construction of public works such as bridges and tunnels, roads and railways, dams and pipelines, airports and docks.
Mechanical engineering	Design, planning, installation and maintenance of plumbing, heating, ventilation, refrigeration and air conditioning, lifts, escalators and fire prevention systems.
Electrical engineering	Design, installation, commissioning and maintenance of electrical supplies and equipment, for lighting, heating, ventilation and communication purposes.
Design	Design of buildings, towns, cities and landscapes, and civil engineering works.
Property management	Maintenance, alteration, repair, restoration and refurbishment of existing buildings. Surveying and property valuation, facilities management.

All of the previous activities which generate and maintain the built environment can be related to activity in the following areas:

- domestic or residential
- commercial
- industrial

Residential construction

Residential construction contributes one-third of Britain's construction output. Types of accommodation built will vary according to need but are usually:

- built to rent from local authorities (20%)
- speculatively or purpose built for owner-occupation (67%)
- built for lease from private landlords or housing associations (13%)

Public sector housing is mainly in the form of **council housing**. Latterly this has decreased as encouragement of individual ownership of properties has increased. This in turn has encouraged the rise of **speculative building**. Housing associations which receive funding from the Government often allocate some of their accommodation to local authority tenants.

Leased residential accommodation from **private landlords** involves the generation of flats and bedsits in older existing properties. This often requires conversion, adaptation or rehabilitation work.

Typical forms of residential accommodation are:

- houses – terraced, semi-detached, detached
- bungalows
- flats
- maisonettes
- bed-sits

Commercial construction

Commercial activity is related to the idea of promoting and delivering goods or services to people who are prepared to pay for them. To allow this to occur various building forms are constructed which will enable:

- shopping activities
- business activities
- leisure activities
- care activities
- spiritual activities
- educational activities

Industrial construction

Industrial activity in the built environment is associated with the production, conversion, assembly, or modification of materials and components prior to their sale. Buildings for these purposes need to possess flexibility with respect to the use of internal spaces and adaptability to allow for alteration and expansion to meet future needs.

Industrial buildings are becoming increasingly more sophisticated as the internal operations and electronic manufacturing plant demand higher building specifications. Staff also, are expecting pleasant, comfortable working environments which produces implications for designers and financial backers.

Speculative development of industrial premises often associated with business or science parks has become an increasingly attractive opportunity for investors. Shell and core units are constructed which contain sufficient flexibility of space, and core services which allow for the building to be customised to suit client requirements.

Organisations in the built environment

It has already been shown that the construction industry has:

- many different and varied clients
- variety in the type of buildings required
- various ways for funding projects
- variety in the type of work that is carried out

This has led to the emergence of many 'organisations' which are capable of responding either in whole or part to those requirements. These organisations may be classed as:

- designers
- local authorities
- large contractors
- small contractors
- sub-contractors

Each organisation is able to function as an entity in its own right, but in addition is able to integrate with one or more of the others to be able to fulfil the requirements of a particular client.

Designers

Design activity brings together a wide range of considerations or individual specialisms which are able to be co-ordinated to produce a development which will meet the technical and functional needs of its final environment. A design team is therefore formed.

In order for designers to work effectively they need to be able to be in contact with varying ranges of people with particular skills and knowledge, a variety of information sources and technical equipment. The team will focus their attention on a particular project or projects depending upon their size and complexity, each with its own resource requirements.

A typical **design team** for a low-rise development consists of:

- **Architects** whose formal training is centred around the activity of designing to meet clients' requirements. Normally the architect will be the leader of the design team.
- **Architectural technicians**, professional **architectural technologists** or **building engineers** who advise on the buildability and performance of different methods of construction, and assist in communicating project information accurately.
- **Structural engineers** who advise on appropriate dimensions and locations of load-bearing elements in the construction.
- **Quantity surveyors** who advise on building costs related to the design, selection of components and amount of human and material resources required for a construction.
- **Service engineers** who advise on appropriate systems for heating and ventilation, electrical and gas installation, plumbing and drainage, communications, security, fire detection and control.

While these people form the design team they may not all be located in the same place. Often they practise from their own premises offering consultancy and professional services to other clients. In this instance they would be contractually bound to honour their professional commitments to individual contracts.

Others who may form part of the team include:

- **surveyors** who provide information about a site or building upon it
- **facilities managers** who will represent the clients' interests more directly
- **landscape architects** who oversee the design and formation of spaces around buildings
- **interior designers** responsible for the co-ordination of the internal environment
- **building contractors** who can advise on the practicalities of theoretical design and buildability aspects.

Local authorities

Local authorities, besides being potential clients for the construction industry, also assist in construction activity by discharging powers delegated to them from central government. This involves the implementation of government policies through statutory instruments. These in turn may allow the development of further legislation known as byelaws.

Local authorities have chief officers who are supported by technicians in order to implement and discharge the duties required of them. Those areas which relate to the built environment are:

- building control
- town and country planning
- environmental health
- highways engineering
- public health
- fire safety
- environmental protection
- safety of existing buildings

Some local authorities also have their own design teams which contain all the specialists previously noted, though some authorities use external consultants to supplement their own design team.

Large building contractors

Large building contractors are firms which are able to have several different projects running at the same time, while separate groups of staff operate from a head office possibly many miles from the construction site. They normally employ over 600 people. There are only a small number of contracting companies that fall into this category.

These are commonly called **main contractors**. They may employ all the specialist labour that they require which is known as the **direct labour force**. Where the main contractor chooses not to do this then specialist labour has to be brought in. This involves the use of either **smaller contractors**, or **sub-contractors**. These other contractors may be **labour only** or **labour and materials**. Contractual agreements are drawn up between the various parties to confirm their individual responsibilities to each other.

Large contractors may be building or civil engineering firms, which operate at local, national or international level and over a wide range of construction activity (Fig. 1.71). The organisation of a medium to large size company is shown in Fig. 1.72.

Characteristics of a large contractor are:

- the ability to respond rapidly to changes in customer requirements
- always seeking to develop more efficient and effective ways of working to generate profit
- the ability to take calculated risks in return for large profit either immediately or in the future

The work range typically covers:

- large housing estates
- multi-storey blocks of flats
- leisure parks
- regeneration schemes, e.g. docklands
- civil engineering (dams, docks, airports, motorways, railways, bridges)
- business parks
- industrial estates
- shopping complexes

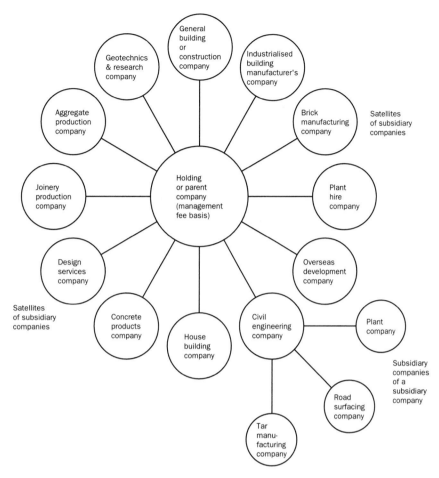

Figure 1.71 Major construction group organisation

Medium size contractors

These employ between eight and 600 people and are engaged in quite large projects but usually operating within a regional location. The majority of companies in this category employ less than 60 people.

The work range typically covers:

- housing
- offices
- warehousing
- industrial units
- refurbishment
- schools
- surgeries
- shops

Small contractors

The majority of companies in the construction industry fall into this category, employing between one and eight people. They are mainly involved in dealing with minor works or very small jobs in a particular locality. This includes the repair and maintenance of buildings. A high proportion of the companies are self employed persons or one-person businesses. They tend to act independently of each other and possibly in competition with each other.

The work range typically covers:

- small housing developments (up to four on site)
- individual residences
- maintenance and repair
- small extensions
- improvements (kitchens, bathrooms, re-roofing, re-wiring)
- central heating installations
- re-decoration

Sub-contractors

Sub-contractors may also be classed as small contractors. The classification of sub-contractor owes more to the fact these companies provide specialist services to small, medium and large companies in order for them to complete their contract.

They may be labour only or labour and materials. On large contracts sub-contractors may be nominated. This means that they have a recognised specialism such as lift installation, refrigeration or fire control. They are appointed by the architect. All other sub-contractors are appointed by the contractor.

The work range typically covers:

- domestic services (water, gas, electric, heating, drainage)
- refrigeration
- transportation (lifts, escalators, travelators)
- suspended ceilings
- scaffolding
- groundwork
- false work

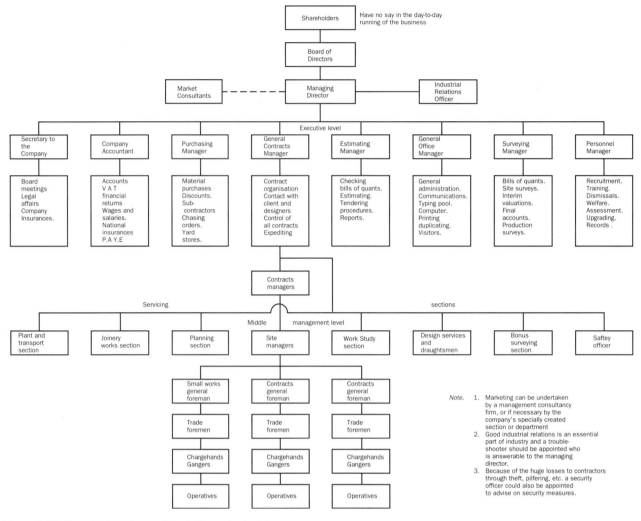

Figure 1.72 Organisation of a medium to large size building company

- trades (brickwork, joinery, plumbing, electrical, decorating, plastering)
- landscaping
- shopfitting
- roofing

Self-assessment tasks

1. List all the different jobs you can think of related to working in the built environment.
2. Group these jobs together under distinct headings.
3. Draw a diagram to show how these jobs might relate to the organisations in the built environment.
4. Describe the typical construction activity that your local council might require.
5. Explain the meaning of the term 'speculative construction' and identify the types of construction it might apply to.

Job functions required by different organisations in the built environment

The development and maintenance of the built environment involves a sequence of operations carried out by different specialists, often overlapping with one another in terms of function in order to provide continuity (Fig. 1.73).

These operations or job functions are:

- **planning** – assessing the suitability of particular sites for particular forms of development
- **designing** – working out ways of organising sites and buildings together, in order for them to meet their owners' requirements
- **producing** – purchasing materials and components, and assembling them on site within specified constraints of time and money, in accordance with the design proposals
- **controlling** – ensuring that each of the previous functions is performed in accordance with defined criteria of economic, social and technical desirability

Planning

This activity revolves around the first decisions to be made about the selection of a site for a particular development. The size and nature of the development will dictate who becomes involved and at what stage.

A site will require consideration of:

- **access** to the surrounding infrastructure – mainly for vehicles and pedestrians

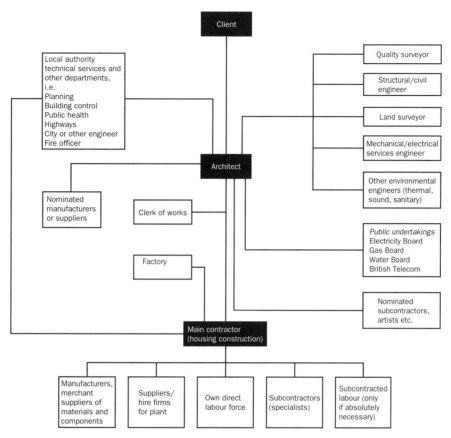

Figure 1.73 Interaction of different job functions in the built environment

- **form** and **character** of the surroundings – the materials and proportions of adjacent buildings, their style and arrangement, and landscaping of the spaces between them
- the **social** and **cultural** context – the geography, history and future of the area, population characteristics, and the public amenities available

This kind of activity is usually undertaken by surveyors or planners. Surveyors specialise in the business and financial aspects of a development, and planners in the handling of political and social issues.

- Surveyors draw upon the knowledge of the property market and the management of land and its associated assets (agriculture, minerals, buildings, and infrastructure) – generally based upon measurable quantities such as prices, dimensions, investment possibilities and financial programming.
- Planners frame and lead discussion about the long term interests of the community, including most aspects of the environmental impact of development.

Surveyors and planners work in both the public and private sector offering services to the public or large commercial organisations, or to local authorities or other statutory bodies.

Designing

The process of designing buildings and their surroundings involves the integration of a wide variety of considerations about the future:

- **durability** – the soundness of a building's construction, based upon the performance of its fabric in changing conditions of loading, temperature, and humidity, both internally and externally
- **usefulness** – the convenience of a building's internal and external layout in relation to the purposes for which it is to be used, changing over time
- **attractiveness** – the continuing pleasure given by a building to people in and around it

Architects are trained to design built environments that combine technical and functional efficiency with consideration for aesthetic values. **Architectural technicians** who support the architect and to some extent surveyors are able to produce designs which demonstrate technical efficiency, which corresponds to functional requirements but often disregards the human dimension.

Specialist design input is mainly from professional mechanical and electrical engineers, structural engineers and quantity surveyors. Occasionally town planners, civil engineers and landscape architects may be involved.

Producing

The production of buildings requires the co-ordination of a wide range of resources:

- deliveries of materials and components, and space in which to store or move them
- employment of relevant skilled and unskilled labour at the right times
- availability of appropriate tools and equipment

- sufficient finance to resource the operations
- managerial expertise to procure and organise projects

Typical functions involved with producing a building:

- **estimating** resources from information provided by designers
- **tendering** in competition with other contractors in order to win a contract
- **contract awarded** and **main contractor** appointed
- **contract planning** decides method and sequence of construction activities
- **buying** the materials and plant required to produce the building
- **setting out** by engineers to determine building position and checking associated operations through the contract
- **cost control** to monitor progress and recover building costs
- **co-ordinating production** on site by a **site agent**
- **construction of the building** by the **operatives** and **sub-contractors**
- **safety planning** by the **planning supervisor**
- **project management** assists in co-ordinating the construction process

Controlling

Most of the control functions related to the built environment are discharged through local authorities under powers delegated from parliamentary legislation. Other bodies such as water authorities, gas and electricity suppliers and telephone companies have statutory powers in the same way as local authorities. They are publicly accountable for their activities and are therefore described as **statutory undertakers**, but operating decisions are taken privately.

The most significant functional areas of control exercised by local authorities in respect of the built environment are:

- **Planning controls** – permission is required for new development, for certain alteration works and demolition work, for cutting down and pruning certain trees, for the display of advertisements, for the storage or disposal of hazardous substances, or for the extraction of minerals and waste disposal (Planning Officer).
- **Building controls** – local authorities are responsible for ensuring that new building work and certain changes of use of buildings comply with the Building Regulations, that dangerous or defective buildings are repaired, or that building structures are demolished safely (Building Control Officer).
- **Fire** and **general safety** – fire certificates are now issued by local authorities, and sports grounds require **safety certificates** (Fire Officer).
- **Environmental health controls** – local authorities are responsible for preventing nuisances or health hazards caused by animals, waste materials (including smoke and fumes), lack of fresh air, noise, overcrowding, or obstructing or damaging highways (Environmental Health Officer).

Other uses of land require **registration** and **licences** issued by the local authority. Usually attached are restrictive conditions applying to notably theatres, cinemas and other places of public entertainment.

Self-assessment tasks

1. Explain the role of the local authority in relation to work on the construction site.
2. Design teams are an important part of the built environment. Describe the role of the design team members that would be involved in planning a new town.
3. The construction of a 1000 property housing estate is to be undertaken by a main contractor. Describe the roles that the contractor's team will play throughout the project.
4. Maintenance activities form a large part of built environment activity. Identify the people who will be involved in the maintenance activities for a national chain store.

Career opportunities in the built environment

Because the production of our built environment is such a large scale and complex operation, the construction industry demands a wide range of knowledge and skills. Traditionally employers recruit people for work in the industry by defining the tasks to be performed and matching them to candidates who present themselves at the interview.

The industry has traditionally compartmentalised itself into three hierarchical layers, namely: operatives, technicians and professionals.

- **Operatives** bring the built environment together physically on site
- **Technicians** communicate information about the construction of the built environment
- **Professionals** take decisions about the formation of the built environment

The industry prides itself on the career progression routes it possesses. It is possible to progress through the industry to professional level provided appropriate experience and qualifications have been achieved. Alternatively it is possible to enter the industry at a level which corresponds to educational attainment, receive appropriate training and experience and consolidate a career from that entry point (Fig. 1.74).

Operatives

Operatives on site are known as skilled operatives or general building operatives (Table 1.8):

- **craft operatives** possess expertise based on years of practice in specific trade areas related to the material or technical system in which they have developed skills
- **general building operatives** generally assist the crafts based operatives by mixing and transporting components, materials and accessories as required, and by clearing and cleaning the site and premises prior to occupation on completion

Table 1.8 Traditional construction crafts

Craft	Activity
Bricklayers	Bed bricks and blocks in mortar to produce walls of many shapes and sizes
Carpenters	Work mainly on-site with timber making, cutting, and fixing both structural and non-structural elements, permanent and temporary (called **formwork**), concealed and decorative
Concretors and floor layers	Spreading ground floor slabs to the correct thickness and levelling screeds
Demolition specialists	Taking down existing construction safely and salvaging materials for re-use
Electricians	Installation of power and lighting circuits, plant and equipment
Joiners/wood machinists	Work off-site producing timber components such as doors, windows, staircases
Painters and decorators	Preparing surfaces and applying paint, varnish, wall-paper, fabrics and sign writing
Plasterers	Mixing and applying plaster and renders, achieving decorative effects
Plumbers	Installation of rainwater goods, boilers and heating systems, gas appliances, hot and cold water systems, sanitary ware, and leadwork
Slaters/tilers	Fix roof coverings and wall claddings
Stone masons	'Bankers' cut and smooth the stones while 'fixers' erect them prepared
Wall and floor tilers	Cutting and laying tiles of different kinds to particular patterns

Latterly there has been a growth in the requirement for **suppliers** to take responsibility for fixing of their own products. This has produced non-traditional skilled trades involving:

- fencing and paving
- scaffolding
- steeple jacks
- plant operators
- furniture installers
- dry lining
- partition fixing
- suspended ceilings
- raised floors
- cladding systems
- thermal insulation

There are some trade bodies which crafts people can become members of, some being:

- Institute of Carpenters
- Guild of Bricklayers
- Institute of Plumbing

Technicians

Technicians (Table 1.9) as previously described are linked with the derivation and communication of information about construction and built environment matters. This communication might be to or from:

- trade operatives
- professional disciplines
- local authority personnel
- materials suppliers

Bodies which technicians may seek membership of:

- British Institute of Architectural Technologists
- Chartered Institute of Building
- Association of Building Engineers
- Board of Incorporated Engineers and Technicians
- Architects and Surveyors Institute
- Society of Surveying Technicians
- Institution of Civil Engineering Surveyors

Professionals

Professional people give good unbiased advice to clients who may be ignorant of technical matters. Decisions are taken and advice is given which has been based on objective criteria and matches best the clients' needs.

Table 1.9 Main technician categories

Technician	Role
Architectural	Supports the architect by interpreting and presenting design information for use by the building team, including other specialists
Building	Site surveying related to setting out of buildings Programming of construction operations Planning and maintaining safety procedures Controlling quality of all site operations Monitoring progress of construction operations Preparing claims for payment Placing orders for materials and plant hire Measuring quantities of work done
Surveying	Assisting in carrying out surveys of land and buildings Developing and maintaining maintenance schedules Advising on preservation or conservation techniques Preparation of plans and specifications Assisting in managing facilities for clients Negotiating with local authorities Preparing building estimates Valuation of property for a client Inspecting work in progress and when completed
Civil engineering	Carrying out site surveys Obtaining and analysing soil samples Developing and producing plans for a project Setting out with lasers Using computer programs to check structural designs Monitoring and recording progress on site Measuring and costing work which has been carried out Co-ordinating safety on site
Building services	Designing and using computerised packages Producing service layouts using CAD Liaison with other construction team members Addressing environmental issues Carrying out energy audits Controlling planned maintenance of services Commissioning new systems

To enable this to be achieved professionals demonstrate their competence by having a background of recognised training and assessment, normally by examination and

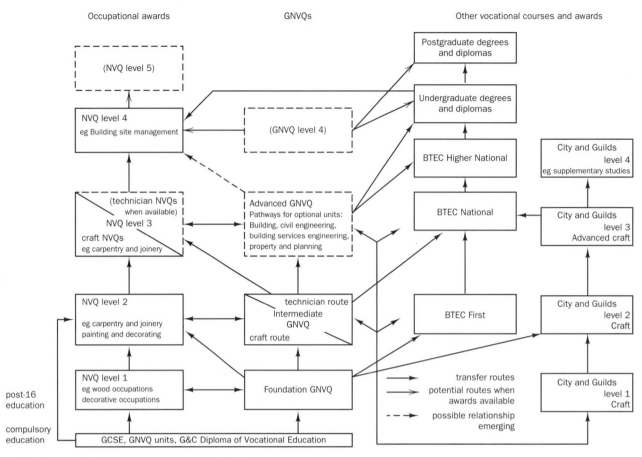

Figure 1.74 Potential progression routes in the built environment

interview before they are admitted to membership of a professional body and allowed to practise.

Construction professionals act as organisers, leaders and co-ordinators of activity. As the industry has developed it has developed areas of specialist knowledge which in turn gave rise to the emergence of the professions that we know today. These professions are:

- **Architecture**
- **Building**
- **Engineering**
- **Facilities Management**
- **Surveying**
- **Town Planning**

Professional bodies associated with these are shown in Table 1.10.

Educational qualifications

There are educational qualifications available for everyone employed in the industry appropriate to their particular type of work. Typical ones are:

Operative qualifications

Normally entered after completing compulsory secondary education.

- City and Guilds Craft and Advanced Craft Certificates
- National Vocational Qualifications at Level 1, 2 or 3 in a trade area

Table 1.10 Professional Bodies

Profession	Professional body
Architecture	Royal Institute of British Architects (RIBA)
Building	The Chartered Institute of Building (CIOB) The Royal Institution of Chartered Surveyors (RICS)
Engineering	Institution of Civil Engineers (ICE) Chartered Institution of Building Service Engineers (CIBSE) Institute of Electrical Engineering (IEE)
Facilities Management	The Chartered Institute of Building (CIOB) British Institute of Facilities Management (BIFM) The Royal Institution of Chartered Surveyors (RICS)
Surveying	The Royal Institution of Chartered Surveyors (RICS) The Chartered Institute of Building (CIOB)
Town Planning	Royal Town Planning Institute (RTPI)

Supervisory qualifications

Normally studied after gaining a craft qualification.

- CIOB First Line Supervisors Certificate
- CIOB Certificate in Site Management Education and Training
- CIOB Diploma in Site Management Education and Training

- CIOB Certificate of Competence in Surveying
- CIOB Certificate of Competence in Computing
- NVQ Level 3 in Supervisory Studies
- NVQ Level 4 in Site Management

Technician Qualifications

Normally entered after completing secondary education with four Grade 'C' GCSEs, or after completing a craft qualification.

- BTEC National Certificate in Building
- BTEC National Certificate in Civil Engineering
- BTEC National Certificate in Building Services Engineering
- BTEC National Diploma in Construction
- BTEC National Certificate in Land Administration

General National Vocational Qualifications (GNVQ)

- Construction and the Built Environment – qualifications at Foundation, Intermediate and Advanced levels

Employment opportunities for GNVQ students are shown in Fig. 1.75. Routes to a professional qualification by part-time study from the GNVQ are shown in Fig. 1.76.

Higher Technician Qualifications

Normally studied after gaining a National Certificate or Diploma, or an Advanced GNVQ. Mature entrants with relevant industrial experience may also be able to enter studies at this level. Entry with one 'A' Level is also possible.

- BTEC Higher National Certificate or Diploma in Building

- BTEC Higher National Certificate or Diploma in Civil Engineering
- BTEC Higher National Certificate in Building Services Engineering
- BTEC Higher National Certificate or Diploma in Land Administration
- BTEC Higher National Certificate in Facilities Management
- BTEC Higher National Certificate in Architectural Technology
- BTEC Higher National Certificate in Planning

Professional qualifications

Professional qualifications require a mixture of academic and professional practice qualifications. The academic qualification may be met in one of two ways: either by the applicant possessing a relevant degree at honours level or by having sat and successfully achieved a pass in the professional body's own examinations.

Relevant professional practice is judged by submitting a report of the candidate's training, experience and responsibilities, and attending a professional interview board. In some institutes a test of professional competence is also a requirement. Success in the interview together with the academic requirement will allow the individual to proceed to full chartered membership of the appropriate body.

It should be noted that education for the professional does not end on attaining professional recognition. It is a requirement that an individual undertakes some form of continuing professional development in order to remain up to date with current practice or knowledge. This is normally recorded in a log book.

Building
- architect/landscape architect
- architectural technician/technologist
- building control officer
- building surveyor
- building technician/technologist
- building maintenance manager/supervisor
- buyer
- clerk of works/inspector
- estimator
- owner/manager
- planner
- quantity surveyor
- site manager/engineer
- site supervisor

Civil engineering
- buyer
- civil engineering technician
- civil engineering/structural design consultant
- clerk of works/inspector
- estimator
- land surveyor
- materials technician/technologist
- planner
- quantity surveyor
- site engineer
- site manager/engineer
- site supervisor/general foreman/ganger

Building services engineering
- building services designer/consultant
- building services surveyor
- building services technician/technologist
- buyer
- clerk of works/inspector
- estimator
- owner/manager
- planner
- quantity surveyor
- site manager/engineer
- site supervisor

Property and planning
- building conservator
- conservation officer
- estate manager
- facilities manager
- housing manager
- land planner
- planner/consultant
- planning technician
- valuer

Figure 1.75 Employment opportunities

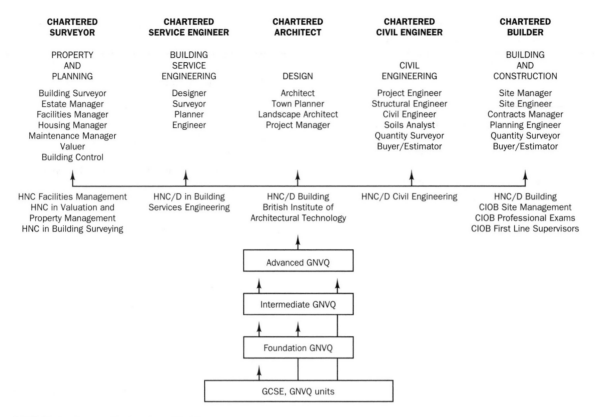

CHARTERED SURVEYOR	CHARTERED SERVICE ENGINEER	CHARTERED ARCHITECT	CHARTERED CIVIL ENGINEER	CHARTERED BUILDER
PROPERTY AND PLANNING	BUILDING SERVICE ENGINEERING	DESIGN	CIVIL ENGINEERING	BUILDING AND CONSTRUCTION
Building Surveyor Estate Manager Facilities Manager Housing Manager Maintenance Manager Valuer Building Control	Designer Surveyor Planner Engineer	Architect Town Planner Landscape Architect Project Manager	Project Engineer Structural Engineer Civil Engineer Soils Analyst Quantity Surveyor Buyer/Estimator	Site Manager Site Engineer Contracts Manager Planning Engineer Quantity Surveyor Buyer/Estimator
HNC Facilities Management HNC in Valuation and Property Management HNC in Building Surveying	HNC/D in Building Services Engineering	HNC/D Building British Institute of Architectural Technology	HNC/D Civil Engineering	HNC/D Building CIOB Site Management CIOB Professional Exams CIOB First Line Supervisors

Advanced GNVQ

Intermediate GNVQ

Foundation GNVQ

GCSE, GNVQ units

Figure 1.76 Routes to a professional qualification

Career progression

It is possible to join the industry at any of three entry points and develop a rewarding career. The student should remember that normally it is not possible to leap-frog a stage in the academic process, as study at one qualification level often is a pre-requisite to the next. It also ensures that the relevant mathematical, scientific and technical knowledge has been adequately developed prior to proceeding to more advanced studies.

Alongside their own academic development students should be aware that it is important to ensure that their practical experience matches the academic level and professional qualification which they seek to attain.

Self-assessment tasks

1. Give six examples of typical activities undertaken by operatives, technicians and professionals.
2. For each of the three categories in question 1 state the academic qualifications they would study.
3. Sketch a typical career route for an operative who wishes to become a Chartered Builder.
4. Explain why operatives might seek to become self employed rather than be directly employed by a main contractor.
5. A student has just graduated from a university and is entering a job related to the built environment. What will the student need to do in order to gain chartered status of a relevant institute?

UNIT 2

The Science of Materials and their Applications

Kemal Ahmet

Welcome to Unit 2! Science is great fun and a very important subject in building. This unit is about understanding the science of materials used in building. Without materials, buildings obviously do not exist. To make safe, reliable, long-lasting buildings it is important to use the correct materials. Clearly, the properties of the materials must be learned and understood. In the modern world there are many different types of materials to use: timber, concrete, plastics, clay products, and as we shall see, many others. It is easy to select the right materials if the appropriate **properties** are known.

In this unit we will be exploring the main **properties** of materials used in construction. We will be learning how some of these properties are **measured**. It is vital to be able to choose the correct materials for a given purpose. Later in the unit, we will be discussing the **selection** of materials for the various parts of buildings.

There are many self-assessment tasks in this module, and the student should try to answer as many of them as possible. Here are some questions to start the reader thinking about materials. Sorry, but no answers are given for these first few problems. The answers will become obvious by the time the unit has been studied!

Problem 1: Figure 2.1 shows part of a bridge. What is the reason for the gap?

Problem 2: What has happened to the timber in the window frame in Fig. 2.2? What has caused the problem?

Problem 3: Why are *light* materials, such as expanded polystyrene, used to keep buildings warm?

Problem 4: Why are the bricks used on top of the wall in Fig. 2.3 different to the others?

Using calculators

Throughout this unit, a calculator will be needed for doing calculations. It is essential that the reader has a scientific

Figure 2.1 A gap, deliberately included during construction, can be clearly seen in this bridge

calculator available while studying this unit. In fact, professional scientists always carry calculators and they do their calculations everywhere! Calculators are very inexpensive to purchase.

Most calculators are very easy to use, but it is always worth reading the manufacturer's instructions. The following quiz should be tried before going on to study the rest of this unit. The reader must ensure that he or she can use a calculator to obtain the correct answer to all the calculations.

Quiz

Try all the calculations with a calculator and make sure that your answers match the ones given below.

Figure 2.2 Damaged window frame

Figure 2.3 A brick wall constructed from two types of brick

1. 8.9×7.6
2. 0.00032×28.3
3. $(45.1 + 50.8 + 47.3 + 49.0) \div 4$
4. $10.1 \div (0.06 \times 0.00085)$
5. $48830 \div 61.6$
6. $\pi \times 12^2 \times 108$

Answers: 1. 67.74; 2. 0.009056; 3. 48.05; 4. 198039 (to the nearest integer); 5. 792.7 (to one decimal place); 6. 48858 (to the nearest integer).

Making estimates

Although using a calculator is very important, the reader should always check that the answers obtained are sensible. There have been some major disasters where engineers have not bothered to check the calculations.

Checks are easy to carry out. The rule is to 'round off' numbers so they are easy to manage. So, checking question 1 above, the figures simply become $9 \times 8 = 72$. This check tells us that the answer is going to be very roughly 72 when the full calculation is carried out.

Checking question 3 above, we can round off the numbers as follows: $(50 + 50 + 50 + 50) \div 4 = 200 \div 4 = 50$. The exact answer, then, is expected to be around 50.

Question 5 above can be rounded as follows: $48000 \div 60 = 4800 \div 6 = 800$. This gives us a rough 'feel' for the correct answer and forms a check. Checks like these help to avoid ridiculous numbers being accepted as the correct answers. Whenever possible, then, the student should make estimates for checking the validity of answers.

Units

Most measurements have units. The base units used in this unit are in Table 2.1.

Table 2.1 Base quantities and their units

Quantity	Unit	Abbreviation
Mass	kilogram	kg
Length	metre	m
Time	second	s
Temperature	degree Celsius	°C

There are other important units, also. As an example millimetres and grams are often used by building professionals. The first self-assessment task should now be done before continuing.

Self-assessment task 2.1

The answers are given at the back of the book. Do not look at the answers until you have tried all the questions!

1. How many millimetres are there in 1 m?
2. Write down two possible units for **area**.
3. Write down a unit for the **volume** of a material.
4. What does **kilo** mean when written in front of a unit?
5. How many grams are there in 1 kg?

Having completed a calculation, the *unit* must not be forgotten. The unit must always be present at the end of a calculation or when a value is written down. For example, if the area of a rectangular room measuring 4 m by 3 m had to be calculated, the answer is *not* 12: this is wrong. The correct answer is $12 \, \text{m}^2$. Be warned!

All the other units used in this book are obtained from those introduced above. Certain combinations of units are named in honour of famous scientists. For example, the unit for force is the **newton** (N). This unit could instead be written as kg m/s^2, but it is much nicer to write this as newton. Likewise, the unit for the rate of heat flow is the **watt** (W) which is the same as kg m^2/s^3. One could use the latter unit if desired but this is both tedious and boring.

Self-assessment task 2.2

Listed below are several units which will be explained in the sections which follow. Write out each unit in full. Remember that the symbol '/' means 'per'. The first one has been done.

Quantity	Unit	Read as
density	kg/m^3	kilogram per metre cubed
Young's modulus	kN/mm^2	
thermal movement	mm/m °C	
thermal conductivity	W/m °C	

2.1 Exploring and measuring the properties of materials

Introduction

In materials science, the properties of materials have to be described. This means that formulae are often needed for calculations. As already explained, a calculator is essential. There are plenty of numerical examples and self-assessment tasks for the student to try. The answers are provided to all the numerical questions.

Students sometimes ask 'Why do we need to do calculations just to describe the properties of materials?' Numerical descriptions are vital for the *correct* scientific description. It is not enough to say that a material is 'strong', or 'heavy', or that it 'feels warm'. Such words or phrases are subjective and may be ambiguous (there may be more than one possible meaning). Verbal description alone is not sufficient in science.

Before going on to describe the properties, the sources of our materials will first be discussed.

Where do materials come from?

Our materials for building can occur **naturally** or they can be **manufactured**. Naturally occurring materials usually need to be **processed** before they can be used. Wood is rarely used directly from trees. Instead, this natural material is first cut into required lengths and then seasoned (dried) to remove the excess moisture. The result is timber. Timber is often treated with preservatives before use, especially if it is to be used on the outside of a building. Likewise, stone needs to be cut to shape and is sometimes polished before being used for building. The more processing that is required, the more costly becomes the material.

Many types of building materials are manufactured. An example of manufactured materials is plastics. These materials are not made by nature, they are *man-made*. Plastics are manufactured chiefly from petroleum. Table 2.2 lists some common materials and the sources of these materials.

When choosing materials, it is important to consider the effect on the environment. Large amounts of energy are required to transport, process and manufacture our materials. Certain materials, for example aluminium, required huge amounts of energy for their production. Although the cost of energy is important it should not be the only consideration. The big problem is that the energy is usually obtained from burning fossil fuels. This means that we are using up non-renewable resources. Further, certain gases which are produced by combustion (e.g. carbon dioxide) contribute to the greenhouse effect. It is believed that this leads to 'global warming'. Various gases produced by combustion produce chemicals which can lead to 'acid rain'. This can cause severe damage to the natural environment. There are many articles in newspapers and magazines on these areas. The student should read some of these articles.

Table 2.2 Some common building materials and main sources

Naturally occurring and processed materials	Source
Timber	Trees in forests or plantations
Aggregates (including sand)	Quarries
Stone (e.g. granite and limestone)	Quarries

Manufactured materials	Main ingredients and source
Timber products (e.g. chipboard, plywood, etc.)	Wood from trees
Clay products (e.g. bricks)	Clay from quarries
Plastics	Petroleum from oil wells
Concrete	Cement manufactured from clay and limestone, plus aggregates from quarries
Metal: steel	Iron from iron ore from mines
Metal: aluminium	Bauxite clay from mines
Glass	Sand from quarries
Bituminous materials (e.g. bitumen and tar)	Petroleum from oil wells

Not only are we unable to replace fossil fuels, but many of the materials used in building are also being exhausted. When local materials have been used up, materials have to be transported in, increasing cost and energy requirements. It should be noted that vegetable-type materials such as timber are in fact **renewable**. With properly managed forests, this material will not run out. Unfortunately, rain forests in various parts of the world are presently being cut down at astonishing rates.

Appearance

The vast range of materials used in the building industry vary tremendously in their appearance. The **texture, reflectance** and **colour** all affect the appearance.

Basically, the **texture** of a material is the arrangement of the patterns on the surface and is affected by the smoothness or irregularities.

Reflectance is a measure of the ability of a surface to reflect light. A perfect reflector of light has a reflectance of 1. A polished mirror-like silver surface has a reflectance of very nearly 1. A plaster surface which has been painted with brilliant white emulsion paint has a reflectance of around 0.8. Surfaces which absorb all the light falling on them have a reflectance of zero.

Quarry tiles have a reflectance of about 0.1. Black velvet cloth has a reflectance of nearly zero. Values of reflectances have been measured and are available for a range of building materials.

The **colour** appearance of materials depends on which colours are reflected back from the surface in the presence of daylight (which is **white** light). White light is a mixture of red, orange, yellow, green, blue and violet light. Thus, a surface appears blue or bluish if it reflects more blue light than the other colours present in white light. Similarly, surfaces which appear red or reddish reflect more red light and less of the other colours which are present. Black surfaces reflect very little light of any colour; most of the light is absorbed.

Appearances vary greatly, even for a given type of building material. A few examples, starting with bricks, are now briefly discussed.

Many types of bricks are available depending on the type of appearance required. As an example, there are in excess of 30 types in the Fletton range of building bricks which are produced by the London Brick Company. The range of colours includes red and buff, and textures range from smooth bricks to those with deep contours. Sometimes new bricks are processed to make them appear to look 'old'. This is achieved by intentionally chipping fragments from the bricks in the factory. These bricks may be used to give a 'traditional feel' to new buildings. Other types of bricks include London stocks which have a characteristic yellow colour and Staffordshire blue engineering bricks, which as the name suggests, have a distinctive blue appearance.

The reader should make a habit of observing the appearance of as many building materials as possible. He or she should not hesitate to look carefully both from a distance and also close up. (The use of a magnifying glass is helpful.) Feeling the materials using hands is also useful. The self-assessment tasks which follow provide some guidance.

Self-assessment tasks 2.3

1. Concrete is commonly used in the manufacture of buildings. It is also used for paths, drives, fencing, etc. Make observations and then describe the appearance of several different finishes of concrete. (There is bound to be plenty of concrete in your neighbourhood!)
2. Make observations and describe the appearance of the following metals:

Material	Typical use (to help you find the material)
lead	flashing on houses
aluminium	frames for replacement windows
copper	pipes and roofing
cast iron	manhole covers

Another category of building material where the appearance varies tremendously is stone. Examples of stones used in building in the UK are granites, sandstones, limestones and slates. A huge range of colours exist:

- granites: mainly pink and grey are produced in the UK (many more colours exist world wide)
- sandstones: typical colours are white, buff, cream, light brown, brown/blue and red, depending on the region of production
- limestones: colours can be creamy white, buff or light brown depending on the type
- slates: colours include black, red, green and brown

The texture of building stones ranges from highly polished (very smooth) to 'rockfaced' (very rough).

Finally in this section, a few words on timber. Timber is one of the most aesthetically pleasing materials available. What makes this material especially interesting is that the appearance of each piece of timber, even from the same tree, is different

from all others. The patterns of the grain on any piece are unique. The colours of timbers vary enormously, also. Certain species are very light in colour while some appear almost black. Texture is variable also, varying from fine to coarse.

Because of the natural beauty of wood, many modern timber preservatives are designed to be *translucent*. These treatments protect the timber but still allow the natural surface texture of the material to be visible.

The reader should make a point of observing a number of different types of timber. Here are some easy to find examples.

Species	Colour	Texture	Where found
Whitewood	White	Fine	Often sold in DIY shops for general purpose use
Beech	Light brown	Fine	Household chairs and furniture
Mahogany	Red/brown	Medium	Window sills, furniture
Scots pine/Redwood	Pinkish	Fine	Main timber used for joinery and cladding

---

Self-assessment tasks 2.4

Unit 2 includes a number of multiple-choice questions to test your knowledge and understanding. For each question, choose the best answer from the options given. Answers are provided at the back of the book.

1. Which **one** of the following materials is naturally occurring?
   (a) steel
   (b) sand
   (c) plywood
   (d) concrete
   (e) PVC
2. Which **one** is a manufactured product?
   (a) polythene
   (b) limestone
   (c) hardwood
   (d) granite
3. Aluminium is produced from
   (a) coal
   (b) crude oil
   (c) bauxite
   (d) sea water
4. The most probable reason why a designer may refuse to specify any South American hardwood is because of the
   (a) low durability of timber
   (b) effect on the natural environment
   (c) poor appearance of this type of timber
   (d) low strength of these materials

Durability

Durability is a measure of the resistance of a material to wear and decay. The slower the deterioration, the greater is the durability. All materials degrade sooner or later. However, a material is said to be durable if it remains in good condition throughout the planned lifetime of a building. The durability depends on many factors such as:

- exposure to sunlight

- presence of water
- frost
- chemical action
- biological factors
- fire

Thus, a material which may be durable when employed indoors may be perishable when exposed to the weather on the exterior of a building. As an example, ordinary chipboard flooring used internally should last the full lifetime of a building. If it is used outside, this material soon degrades.

The term durability should be used in relation to the intended use. Thus, in the example above, chipboard is durable when used in dry conditions. It is definitely not durable when used in the external environment. As another example, bricks have a tendency to absorb water (this is because of the porosity. In cold weather, the absorbed water freezes. If freezing occurs on a number of occasions certain bricks start to deteriorate – they begin to crumble away. This means that these kinds of bricks are not durable in climates where there is damp and frost.

Self-assessment tasks 2.5

State whether you think the following materials are durable for the usage stated:

1. concrete for drives
2. plaster for rendering an external wall
3. clay tiles for roofs
4. plastics for fire doors

Density

Most people have been asked the question: 'Which weighs more: one kilogram of lead or one kilogram of feathers?' Needless to say, they weigh exactly the same: each has a mass of 1 kg. So, what *is* the difference between these two materials? A mass of 1 kg of lead has a much smaller volume than 1 kg of feathers. Lead is said to have high **density**. Feathers have very low density (this helps birds to fly).

Figure 2.4 shows a small cylinder of lead and a block of expanded polystyrene, both having the *same* mass. Both of these materials are commonly used in building. The volume of the expanded polystyrene is about 550 times greater than the volume of the lead. For this reason, the lead is 550 times denser than the expanded polystyrene.

How is the density of any given material calculated? The two quantities always required to find the density are:

- mass (in kg)
- volume (in m^3)

These quantities can easily be measured. When the mass and volume are known, the following formula is used:

$$\text{density} = \frac{\text{mass}}{\text{volume}}$$

In words, the density is equal to the mass divided by the volume. Clearly, the unit for density is kg/m^3, or kilogram per metre cubed. Table 2.3 lists the densities of various materials commonly used in building. The reader should browse through the list and get a 'feel' for the figures. It is worth remembering that the density of pure water is $1000 \, kg/m^3$, or $1 \, m^3$ of water has a mass of 1 tonne. As can be seen from Table 2.3, some of our building materials have very high density while others are very low in density.

Table 2.3 Density of common building materials

Material	Bulk density (kg/m^3)
Expanded ebonite	64
Expanded polystyrene	16–24
Foamed polyurethane	24–40
Glass fibre quilt	16–48
Mineral and slag wools	48
Wool, hair and jute fibre felts	120
Corkboard (baked)	128
Balsa	160
Sprayed insulating coatings	80–240
Insulating fibre building boards	240–350
Exfoliated vermiculite (loose)	80–144
Rigid foamed glass slabs	128–136
Medium fibre building boards	350–800
Aerated concrete (low density)	320–700
Compressed straw slabs	365
Wood–wool slabs	450
Exfoliated vermiculite concrete	400–800
Expanded clay – loose	320–1040
Standard and tempered hardboards	800 and 961 (min)
Softwoods and plywoods	513
Diatomaceous earth brick	721
Asbestos-silica-lime insulating board (BS 3536)	881 (max)
Particle boards	449–800
Plasterboard	961
Hardwoods	769
Exfoliated vermiculite plaster	641
Perspex	1190
Foamed blastfurnace slag concrete	960–2000
Expanded clay and sintered PFA concretes	720–1760
Polyester glass fibre laminate (GRP)	1620
Asbestos cement (semi-compressed (BS 690))	1200
Clinker concrete	1041–1522
Plaster (dense)	1442
No-fines concrete	1142–1842
Asbestos cement (fully compressed (BS 4036))	1600 (min)
Aerated concrete (high density)	1602
Brickwork	1700
Mastic asphalt	2100
Cement : sand	2306
Glass	2520
Rendering	1778
Sandstone	2500
Concrete 1 cement : 2 coarse aggregate : 4 sand	2260
Limestone	2310
Slate	2590
Granite	2662
Zinc	7140
Steel	7850
Aluminium and alloys	2700
Copper	9000
Lead	11 340

Source: Everett *Materials* Longman

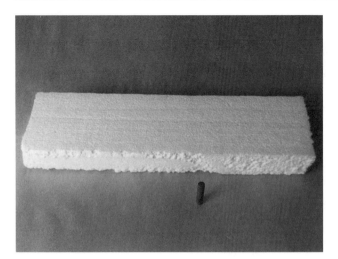

Figure 2.4 Both the lead and the expanded polystyrene shown have the same mass

Before going on to the next section, there is a little more to understand about density. Consider the following. A jar is filled with sand. What happens to the level of the sand if the jar is then gently tapped (without spilling any)? The level goes down. There are millions of small pockets of air between the grains of sand. By tapping the jar, some of the air gaps are filled up by the sand. For this reason the contents settle and the volume of the sand is reduced. As the mass has not changed, the overall density has *increased*. The density, then, depends on how much air is trapped between the particles of solid material. (The air trapped is also known as **voids**.) There are two definitions of density:

- **Bulk density**: this is the overall density of a material. The volume considered includes any air which may be present in the gaps between the particles.
- **Solid density**: this is the density *only* of the solid material. The volume of the air present is subtracted from the overall volume.

Note that in Table 2.3 the *bulk* densities have been provided.

Example

A piece of mahogany to be used for a window sill has a length 2 m, width 0.30 m and thickness 0.025 m. The mass is 12 kg. Determine the density of mahogany.

Answer

$$Mass = 12\,kg$$

$$Volume = 2 \times 0.30 \times 0.025 = 0.015\,m^3$$

So, using the formula for density:

$$Density = 12 \div 0.015 = 800\,kg/m^3$$

Example

A certain quantity of soft sand to be used to make mortar was analysed as follows:

Mass $= 40\,kg$

Overall volume $= 0.025\,m^3$ (this is also known as the *bulk* volume)

Volume of (air) voids $= 0.009\,m^3$

Calculate (i) the bulk density and (ii) the solid density.

Answer

(i) Bulk density $= 40 \div 0.025 = 1600\,kg/m^3$

(ii) To calculate the solid density, the solid volume is needed. This is given by: $0.025\,m^3 - 0.009\,m^3 = 0.016\,m^3$.

So, solid density $= 40 \div 0.016 = 2500\,kg/m^3$

The reader should check these calculations.

Self-assessment tasks 2.6

1. A lightweight concrete building block has sides 0.44 m × 0.215 m × 0.1 m and a mass of 10 kg. Find the density of the block. How does the density compare with the value for normal density concrete given in Table 2.3? Give two advantages of using *lightweight* blocks in building.
2. A student carried out an experiment to measure the density of several building materials. She wrote all the values of mass and volume into a table of results, see below. Use a calculator to complete the last column. Now compare all values with Table 2.3. Are there any values of density you disagree with? Why?

Material	Mass (kg)	Volume (m³)	Density (kg/m³)
Glass	0.75	0.0003	2500
Steel	3.90	0.0005	
Granite	6.75	0.0025	
Foamed polyurethane	3.00	0.1	
Common brick	2.38	0.014	
Plywood	5.60	0.01	

3. A certain type of brick has a mass of 2.52 kg and volume 0.0014 m³. Its density, in kg/m³, is
   (a) 0.000 556
   (b) 0.003 528
   (c) 180
   (d) 1800
   (e) 18 000

Self-assessment tasks 2.7

1. Expanded polystyrene typically has density equal to 20 kg/m³. Calculate the solid density given that 98% of the bulk volume of expanded polystyrene is taken up by air. (Hints: 1 m³ of expanded polystyrene has mass 20 kg. The volume taken up by the solid material is $2 \times 1/100\,m^3$.)
2. Figure 2.5 shows a 3 kg brick next to 3 kg of mineral wool loft insulation. Which material has greater density? Why?

Measure of density

As already explained, for a given material both the mass and the volume are needed in order to calculate its density. The mass is easy to measure with an electronic balance. Figure 2.6 shows a block of concrete being weighed on a balance. It has a mass of very nearly 7.1 kg.

Figure 2.5 The two materials shown each have 3 kg of mass

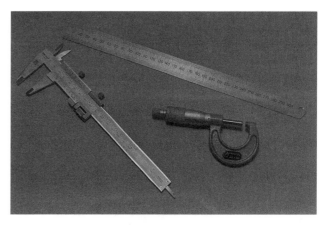

Figure 2.7 A ruler, vernier calipers or micrometer screw gauge can be used to measure the dimensions of a material. From this, the volume can be found

For objects of known shapes, e.g. rectangular and cylindrical shapes, the volume can be found by calculation, once the required dimensions have been measured. For the measurements, a ruler or vernier calipers or a micrometer screw gauge can be used. A ruler is adequate for measuring large sizes, for example the sides of a block or brick can be accurately measured with a ruler. When small distances are to be measured, vernier calipers or a micrometer screw gauge must be used (see Fig. 2.7). An example is when the thickness of a sheet of glass needs to be measured.

For irregular-shaped objects, the volume must be found by other methods. Examples of irregular-shaped materials are the stones used in aggregate. The most effective way of finding the volume of irregular objects is by the **displacement of water** method. Figure 2.8 shows a sample both before and after lowering into a measuring cylinder containing 100 ml of water. The volume of the sample is the difference between the final and original level of water.

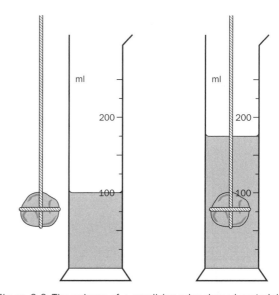

Figure 2.8 The volume of a small irregular-shaped material may be found by immersing it in water in a measuring cylinder

In the example of Fig. 2.8, this is just 175 ml − 100 ml = **75 ml**. (N.B. 1000 ml = 1 litre and 1000 litre = 1 m^3.)

Self-assessment task 2.8

A building scientist needed to find the density of blast furnace slag (which is used to manufacture certain types of concrete). Having found the mass of a certain amount, she tried to find the volume using displacement of water. Figure 2.9 shows the problem she encountered. How can this problem be overcome? (The answer to this problem follows shortly.)

For large objects a **displacement can** (Fig. 2.10) may be used to determine volume. The water level should initially be just at the overflow. The object is tied by a fine thread and is carefully lowered into the container until it is fully submerged. The displaced water is collected using a measuring cylinder (see Fig. 2.10). This gives the volume. The density is then calculated as usual by dividing the mass by the volume.

Figure 2.6 Electronic balance used to measure the mass of a concrete block

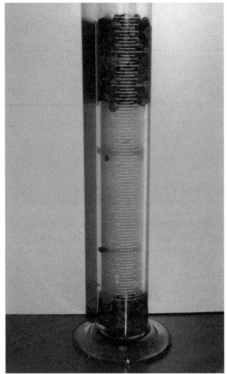

Figure 2.9 The material shown has variable density such that some of it floats while some particles sink

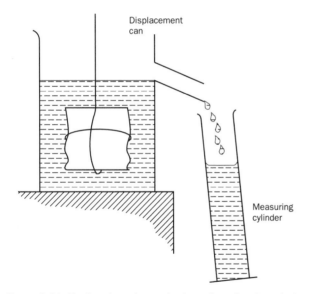

Figure 2.10 Finding the volume of a large irregular-shaped piece of material

If the material is less dense than water it will float. In this case the object should be forced below the water by placing (say) a weight on top of it. The volume of water collected will then be from both the object and the weight

pushing it down (call this V_{total}). The volume of the weight (V_{weight}) on its own is found by using the same method. The volume of the material is then given by $V_{object} = V_{total} - V_{weight}$.

Making estimates: an example

As discussed earlier, it is very important to make estimates. Remember that to estimate means to quickly obtain an answer which is roughly correct, without having to do the entire set of calculations or measurements. Numbers are often rounded-off, for example 9.8 becomes 10, 3.142 becomes 3, etc. Where actual values are unknown, sensible guesses can be made. For example, the height of a room in a domestic property can sensibly be estimated to be 2 m or 2.5 m or 3 m. However, guesses of 1 m or 10 m are obviously inappropriate.

As an example of an estimating type of problem, Fig. 2.11 shows a small brick wall. Estimate the mass of the wall in kilograms.

At first this seems like a very difficult question. By quick arithmetic and some estimates we can obtain a very reasonable answer. Here is a sensible way of tackling this problem. (The values below were obtained both by using Table 2.3 and from approximations. The symbol '≈' means approximately.)

> number of bricks vertically = 10 (obtained from diagram)
> height of one brick including the layer of mortar above
> ≈ 70 mm = 0.07 m (estimated)
> ∴ total height of wall ≈ 10 × 0.07 = 0.7 m
> number of bricks horizontally = 28 (obtained from diagram)
> length of one brick ≈ 220 mm = 0.22 mm (estimated)
> ∴ total length of wall ≈ 28 × 0.22 = 6.16 m
> thickness of wall ≈ 100 mm = 0.1 m (estimated)
> ∴ volume ≈ 0.7 × 6.16 × 0.1 = 0.43 m³

It is now assumed that the brick and mortar have the same density. From Table 2.3, density ≈ 1700 kg/m³. Rearranging the formula for density:

$$mass = density \times volume \approx 1700 \times 0.43 = 731\,kg$$

Therefore, the estimated mass of the wall is around 730 kg. This is about $\frac{3}{4}$ tonne.

Self-assessment task 2.9

Estimate the mass of a wooden garden hut of length 2 m, width 1.5 m and height 2 m. Assume that the roof is flat and take the timber to be 5 mm thick. Obtain an estimate for the density from Table 2.3.

Figure 2.11 Brick wall

Self-assessment task 2.10

The data below shows measurements made using four pieces of timber from different species:

Species	Mass (g)	Length (mm)	Width (mm)	Height (mm)
Cedar	70.5	130	70	20
Hemlock	112.0	145	100	14
Beech	156.8	300	30	25
Greenheart	91.1	115	70	12

Using appropriate software on a computer, enter this information onto a spreadsheet. First, process the data so that the masses are converted to kg and the distances to m. Second, use the spreadsheet to calculate the volume in m^3 and density in kg/m^3 for each material.

Figure 2.12 shows one way of completing the spreadsheet (Microsoft Excel was used in this example). Once the spreadsheet has been completed, it should then be printed out. Calculations should be checked and confirmed using a calculator. The spreadsheet should be saved to a floppy disk so that it can be used for processing other data also.

Thermal properties

To start, here are some important scientific facts:

- the word thermal refers to **heat**
- heat is a form of **energy** and is measured using the unit called **joules**
- temperature measures the hotness of a material and the unit for temperature is °C
- temperature must *not* be confused with heat, which measures energy *flow* into or out of something
- heat always tries to flow from a higher temperature to a lower temperature

Self-assessment tasks 2.11

Remembering the important points from above, you should state the direction of heat flow for each case:

1. Hot summer's day: outside air temperature = 25 °C, inside room temperature = 20 °C
2. Warm spring day: outside air temperature = 20 °C, inside room temperature = 20 °C
3. Cold winter's day: outside air temperature = −5 °C, inside room temperature = 15 °C

Thermal **power** is an extremely important quantity. If the power loss from a building is known, the total energy transferred and the cost involved can be calculated. Note that:

- power is the heat transfer per second
- power is measured in joule per second (J/s) or watts (W): 1 J/s = 1 W

Your own body is losing power at the rate of about 150 W as you read this book. This means that 150 J of thermal energy is transferred to the surroundings *every second*. A typical three bedroom house can easily give off 10 000 W (10 kW) of thermal power on a cold day. This means that 10 000 J of energy is lost by the house to the surrounding environment *every second*.

Thermal insulation

In countries such as the United Kingdom, the outside air temperature is often too low for comfort. Typically, winter temperatures lie between 0 °C and 10 °C. Inside buildings, people require a temperature of about 20 °C to feel comfortable. Buildings must therefore be heated. This means gas, oil, electricity or other fuels must be used. All

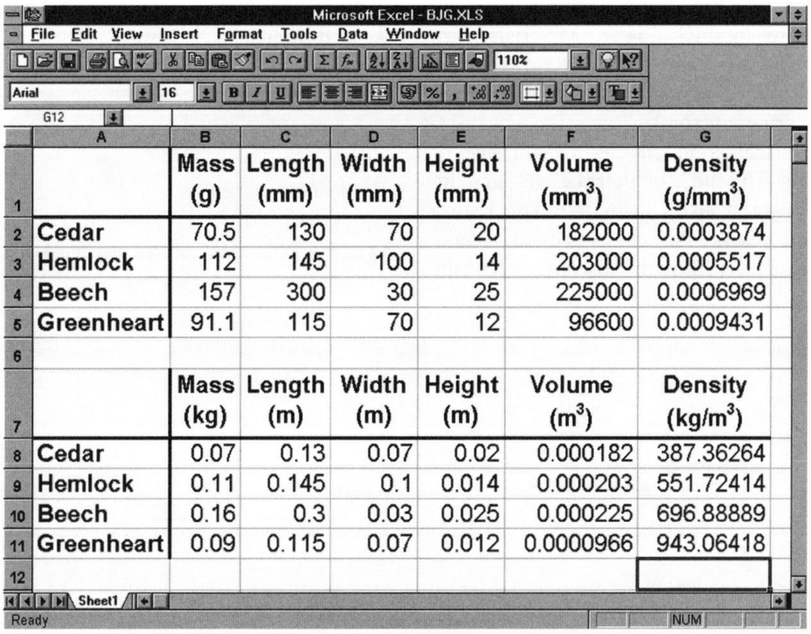

	A	B	C	D	E	F	G
1		Mass (g)	Length (mm)	Width (mm)	Height (mm)	Volume (mm³)	Density (g/mm³)
2	Cedar	70.5	130	70	20	182000	0.0003874
3	Hemlock	112	145	100	14	203000	0.0005517
4	Beech	157	300	30	25	225000	0.0006969
5	Greenheart	91.1	115	70	12	96600	0.0009431
6							
7		Mass (kg)	Length (m)	Width (m)	Height (m)	Volume (m³)	Density (kg/m³)
8	Cedar	0.07	0.13	0.07	0.02	0.000182	387.36264
9	Hemlock	0.11	0.145	0.1	0.014	0.000203	551.72414
10	Beech	0.16	0.3	0.03	0.025	0.000225	696.88889
11	Greenheart	0.09	0.115	0.07	0.012	0.0000966	943.06418
12							

Figure 2.12 A spreadsheet generated for calculating the density of rectangular-shaped materials. The lower table contains the required answers

these forms of energy are expensive. For this reason, it is very important to try to stop as much heat escaping from the internal environment as possible. In fact, the Building Regulations (Part L) require that all new buildings are thermally insulated to at least the minimum values stipulated.

Self-assessment task 2.12

Apart from cost, can you think of other reasons for wanting to reduce the fuel consumption?
(Hint: think of possible damage to the environment.)

Heat is lost from buildings by a number of processes:

- **conduction**: this is heat which flows *through* the outer walls, floors, roof, etc.
- **convection**: this is heat which is carried away by the circulation of gases such as air
- **radiation**: this is given off from the outer surfaces of the building and is known as *infra-red waves*
- **evaporation**: when water on the outer surfaces evaporates, it carries away heat

In buildings, conduction is the most important method by which heat is transferred. In fact, conduction is the only method which allows heat to flow through the solid material of the building fabric. For this reason, conduction is now discussed in more detail.

Figure 2.13 represents an outer wall. Clearly, the thermal power loss by conduction depends on:

- **area** of the wall: as the area increases the power loss increases
- **thickness** of the wall: as the wall becomes thicker, the power loss *decreases*
- **temperature difference** between the inner and outer *surfaces* of the material: as the difference in temperature increases, the power loss increases
- **thermal conductivity**: this is discussed below

The thermal conductivity measures the ability of a given material to conduct heat. Materials with a greater thermal conductivity transfer more thermal power. Materials with high thermal conductivity are said to be good thermal **conductors**. Materials with low thermal conductivity are good thermal **insulators** (or poor conductors).

Table 2.4 lists the thermal conductivities of a range of materials used in building. The unit for thermal conductivity is watts per metre per degree Celsius (W/m °C). The reason for this will be explained below. The reader should browse through the list to get a 'feel' for the numbers. As one

Table 2.4 Thermal conductivity of various building materials

Material	Thermal conductivity (W/m °C)
Expanded ebonite	0.029
Expanded polystyrene	0.035; 0.033
Foamed polyurethane	0.024; 0.039
Glass fibre quilt	0.032–0.04
Mineral and slag wools	0.03–0.04
Wool, hair and jute fibre felts	0.036
Corkboard (baked)	0.040
Balsa	0.045
Sprayed insulating coatings	0.043–0.058
Insulating fibre building boards	0.053–0.058
Exfoliated vermiculite (loose)	0.047–0.058
Rigid foamed glass slabs	0.050–0.052
Medium fibre building boards	0.072–0.101
Aerated concrete (low density)	0.084–0.18
Compressed straw slabs	0.101
Wood-wool slabs	0.093
Exfoliated vermiculite concrete	0.094–0.260
Expanded clay – loose	0.12
Standard and tempered hardboards	0.125 and 0.180
Softwoods and plywoods	0.124
Diatomaceous earth brick	0.141
Asbestos-silica-lime insulating board (BS 3536)	0.144 (max)
Particle boards	0.101–0.158
Plasterboard	0.16
Hardwoods	0.16
Exfoliated vermiculite plaster	0.19
Perspex (ICI)	0.21
Foamed blastfurnace slag concrete	0.24–0.93
Expanded clay and sintered PFA concretes	0.24–0.91
Polyester glass fibre laminate (GRP)	0.35
Asbestos cement (semi-compressed (BS690))	0.37
Clinker concrete	0.37–0.58
Plaster (dense)	0.48
No-fines concrete	0.562–0.75
Asbestos cement (fully compressed (BS 4036))	0.65
Aerated concrete (high density)	0.65
Brickwork	1.45–0.73
Mastic asphalt	0.60
Cement:sand	0.53
Glass	1.05
Rendering	1.15–1.21
Sandstone	1.29
Concrete 1 cement : 2 coarse aggregate : 4 sand	1.44
Limestone	1.53
Slate	1.88
Granite	2.93
Zinc	117.64
Steel	57
Aluminium and alloys	214
Copper	400
Lead	35.71

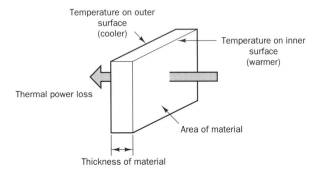

Figure 2.13 Diagram showing power transfer through a wall

example, copper has a thermal conductivity of 400 W/m °C, while expanded polystyrene has a value of 0.033 W/m °C. The expanded polystyrene is about 12 000 times better at insulating heat than copper!

Materials with many small air pockets are usually good insulators. Here are some examples. Foamed polyurethane and expanded polystyrene are often used in wall insulation. Glass-fibre quilt is used to insulate lofts, reducing heat loss from the roof (see Fig. 2.48). All these materials have one thing in common. They contain substantial amounts of air, which is an excellent thermal insulator. Sometimes students ask: 'if air is such a good thermal insulator, why do we fill cavity walls (which originally contained *air*) with foam?' The answer to this question is that free air can transfer heat efficiently by *convection* currents. Air must be trapped in small pockets to ensure that convection does not take place. In this way the thermal insulating ability of the air is fully utilised.

So far, the discussion of thermal insulation has been mainly qualitative (without calculations). To work out the power loss which results from conduction, the following formula is used:

$$\text{power} = \frac{\text{thermal conductivity} \times \text{area} \times \text{temperature difference}}{\text{thickness}}$$

This formula can only be used when the temperatures on both the inner and outer surfaces of a material are stable, that is, in equilibrium.

The formula above can be rearranged to make thermal conductivity the subject:

$$\text{thermal conductivity} = \frac{\text{power} \times \text{thickness}}{\text{area} \times \text{temperature difference}}$$

(The reader should check that this is correct.) If the units on the right-hand side of the equation are written out, the unit for thermal conductivity becomes:

$$\frac{\text{W} \times \text{m}}{\text{m}^2 \times {}^\circ\text{C}} = \frac{\text{W}}{\text{m} \, {}^\circ\text{C}} = \text{W/m} \, {}^\circ\text{C}$$

To make sure that the correct answer is obtained, these units *must* be used every time.

How can the thermal conductivity of a material be measured? This is normally done by scientists in a laboratory. They use the formula above having measured the following quantities:

- area of one face
- thickness of the material
- power transfer
- temperature difference between the inner and outer surface

The next self-assessment task tests the ability of the reader to use the formula for the thermal conductivity.

Thermal movement

An increase in temperature causes most materials to **expand** (swell). Likewise, when the temperature is lowered, most materials **contract** (shrink). In building, the general name given to this effect is **thermal movement**. The amount of expansion or contraction is usually too small to be noticeable by eye. However, the forces involved in expansion are extremely large. For this reason, allowances must be made to ensure that parts of buildings can expand or contract freely. Expansion gaps allow this to happen; Fig. 2.14(a) shows an example.

Figure 2.14(b) shows the damage caused to the side of the building by the expansion of a long brick wall. The

(a)

(b)

Figure 2.14 (a) Expansion gap in a large building; (b) damage caused by expansion of a brick wall

Table 2.5 The coefficient of thermal expansion for various materials.

	Material	Coefficient of thermal expansion ×0.001* (mm/m °C)
High expansion	Polythene, HD/LD	144/198
	Acrylics	72–90
	uPVC	70
	Timber–across fibres	30–70
	Phenolics	15.3–45
	Zinc	31
	Lead	29
Medium expansion	Aluminium	24
	Polyesters	18–25
	Brass	18
	Copper	17.3
	Stainless steel	17.3
	Gypsum plaster	16.6
	GRC	13–20
Low expansion	Sandstones	7–16
	Concretes – various aggregates	10–14
	Mild steel	11–13
	Glass	6–9
	Granite	8–10
	Slates	6–10
	Marbles	1.4–11
	Limestones	2.4–9
	Aerated concretes	8
	Plywood	4–16
	Fire clay bricks	
	length	4–8
	width and height	8–12
	Mortars	11–13
	Asbestos-cement BS 690	12
	Asbestos-silica-lime insulating board	5
	Timber–longitudinal	3–6

*Note that all figures must be multiplied by 0.001 as indicated at the top of the third column

expansion occurred during a hot summer. The builders provided no expansion gaps in the long wall. There was also no expansion gap between the wall and the building. In fact, a similar effect occurred at the other end of the wall, damaging another part of the building. This is clear evidence that when expansion occurs, the forces involved are extremely large.

The amount of expansion varies from one material to another. This ability to expand (or contract) is measured using the **coefficient of thermal expansion**. Table 2.5 lists the coefficient of expansion for a number of building materials. Remember that the bigger the value, the more the expansion for an increase of 1 °C in temperature. Consider aluminium. The coefficient of thermal expansion is 0.024 mm/m °C. This means that 1 m of aluminium will increase in length by 0.024 mm when the temperature rises by 1 °C. The length *decreases* by this amount when the temperature drops by 1 °C.

The thermal movement can be calculated as follows:

change in length = coefficient of thermal expansion ×
change in temperature × original length

The following example uses this formula. The student should work through it to check the calculations.

Before doing the next self-assessment task, remember that to calculate the thermal movement the following must be known:

- coefficient of linear expansion (e.g. from Table 2.5)
- change in temperature
- original length

Example

Calculate the increase in length of a steel beam of length 10 m if its temperature rises from 20 °C to 400 °C during a fire.

Answer

From Table 2.5,
coefficient of thermal expansion = 0.012 mm/m °C
(average of range given)
change in temperature = 400 °C − 20 °C = 380 °C
original length = 10 m

Using the formula:

change in length = 0.012 × 380 × 10 = 45.6 mm ≈ 46 mm

This change in length causes severe warping and buckling of the beam.

Self-assessment tasks 2.15

1. A uPVC waste disposal pipe has length 3.5 m and is used to dispose both cold and hot liquids. Cold liquids which pass through are typically at 10 °C and hot liquids can be up to around 70 °C. Estimate the maximum change in length which can take place. (Note that allowances must be made during installation to allow such pipes to expand and contract.)

2. A material which suffers from high thermal movement
   (a) is a good thermal conductor
   (b) hardens as it is heated
   (c) expands a lot as the temperature rises
   (d) burns easily

3. The coefficient of linear expansion for a certain building material is 0.017 mm/m °C. By how much will a 4 m length expand if its temperature rises by 30 °C?
   (a) 0.002 mm
   (b) 0.13 mm
   (c) 2.04 mm
   (d) 7059 mm

4. An experiment was carried out to measure the coefficient of expansion for a certain plastic. The (original) length of the plastic was 1000.0 mm at 20 °C. The following lengths were recorded at the temperatures shown:

Length/mm	1000.0	1001.5	1003.1	1004.4	1006.0
Temperature/°C	20	30	40	50	60

   Carry out the following tasks:
   (i) Plot a graph of *increase* in length (*y*-axis) against *increase* in temperature (*x*-axis). Make sure the axes are correctly labelled and include the correct units.
   (ii) Draw the best straight line through the points, ensuring that the line passes through the origin. Explain the main features of the graph.
   (iii) Calculate the gradient of the line, and ensure that you include the appropriate unit.
   (iv) The gradient you have measured is the coefficient of thermal expansion (because the starting length was exactly 1 m). Using Table 2.5, decide the type of plastic which may have been used in the experiment.

Porosity

Many building materials are **porous**. This means that they are full of **pores**. Pores are very tiny holes within the material. They form very small openings on the surfaces. Water may pass into and through the material by travelling through the pores. The process which allows the water to be absorbed is called **capillary action**. Bricks, concrete and timber are examples of porous materials. On the other hand, a range of materials are not porous. Examples of *non*-porous materials are

● mastic asphalt: this is used for water-proofing flat roofs
● aluminium: thin sheets are used to stop water vapour passing through
● polyvinyl chloride, PVC: used for damp-proof courses

The importance of non-porous materials is discussed in the next section.

Before reading further, the reader should try the following simple experiment. This will enable him or her to understand porosity and capillary action. One sheet of blotting paper or kitchen roll should be held vertically over a bowl of water.

The paper is then lowered until one side just touches the surface of the water. Observations will show that the water rises up. The water is efficiently drawn into the material because capillary action 'sucks' the water into the pores.

Are there problems in using porous building materials? As we have seen, porous materials are able to absorb (or lose) moisture quite efficiently. A change in the moisture content of a material often causes 'movement' (swelling or shrinking). This is discussed further, below. Rising damp is caused by the porosity of materials. Also, if a material absorbs moisture which then freezes (in cold weather), the material can then become damaged. This is because water expands as it freezes. Figure 2.15 shows frost damage to bricks. Another problem when moisture absorption occurs in porous materials is that the thermal conductivity increases. This means that the thermal insulation is reduced. For various reasons, then, it is necessary to be able to calculate the porosity of a material.

The porosity is defined as follows:

$$\text{porosity} = 100 \times \text{volume of pores} \div \text{bulk volume}$$

Note that bulk volume refers to the **overall** volume of the material (see also the section on density). A material with no pores has a porosity of 0%. Suppose that one-quarter of a certain material is full of air (inside the pores). In this case, the porosity is simply 25%.

There is another formula for calculating the porosity of a material:

$$\text{porosity} = 100 \times \frac{\text{solid density} - \text{bulk density}}{\text{solid density}}$$

Figure 2.15 The porosity enables moisture to enter the bricks. Cold weather causes the moisture to freeze. After a number of cycles of freezing and thawing, frost damage begins to become apparent.

Example

A lightweight concrete block has solid density 2400 kg/m^3 and a bulk density of 800 kg/m^3. Calculate the porosity.

Answer

This is easy to do by substituting straight into the second formula:

porosity $= 100 \times (2400 - 800) \div 2400$

$\qquad = 160\,000 \div 2400 = 67\%$

In other words, two-thirds of this material contains air inside the pores.

Self-assessment task 2.16

Calculate the porosity of a certain common brick given the following information:
Bulk volume $= 1400$ ml
Volume of the solid material (i.e. excluding the volume of the pores) $= 770$ ml

---

Moisture barriers

Before reading further, the following self-assessment task should first be tried.

Self-assessment task 2.17

In the section above, several non-porous building materials were mentioned. Can you list them? (Try not to look!)

Non-porous materials do not absorb or transmit moisture. These materials have various uses in building. Some of these are briefly discussed:

- **Moisture barrier.** This is a general term having the same meaning as **damp proof membrane** (dpm). Sheets of bituminous felt or plastic (e.g. polyvinylchloride) are used in pitched roofs as sarking. The moisture barrier prevents any water which has penetrated the roof tiles from getting into the roof space. Likewise, rising damp is eliminated on solid ground floors by a continuous sheet of plastic.
- **Damp-proof course** (dpc). Bricks and blocks are porous. Without a dpc, the outer walls of houses can suck up moisture from the ground causing damp problems. The damp damages internal surfaces and causes timber to rot. The presence of damp also encourages various diseases. For this reason a dpc is essential. Historically, non-porous materials such as slate or lead or copper were used. Nowadays, a strip of bituminous felt or PVC is present in the wall, around 150 mm above ground level. These materials are not porous and so capillary action cannot take place. This ensures that moisture remains safely below the dpc.
- **Vapour barrier.** Washing, cooking and people breathing all produce moisture in the form of water vapour (a gas). When moist air hits a cold surface, the water vapour condenses – or turns back into liquid water. We are all familiar with water forming on window panes on a cold day. Moist air can penetrate through plaster, blocks and bricks because these are porous materials. Water vapour

which passes from inside a house, into a wall, and then condenses out causes *interstitial* condensation. This effect can cause long-term damage to the structure. Further, materials which become damp also become less efficient at heat insulation. To avoid the possibility of interstitial condensation, a **vapour barrier** is installed. This is a layer of material which resists the passage of water. As already discussed, these are non-porous materials. Examples of vapour barriers include aluminium foil, polyethylene sheeting and bituminous felt.

Moisture movement

The reader should think about the answer to the next self-assessment task before going any further.

Self-assessment task 2.18

External timber doors often 'stick' during wet weather, making them difficult to open or close. What causes this effect? (If you are not sure, the answer will become clear in the next paragraph.)

In an earlier section, we saw that movement occurs when the temperature changes. Similarly, with many materials, movement occurs when the moisture content changes. In fact, the answer to the self-assessment task above is that swelling (expansion) occurs when the timber absorbs more moisture. This is an example of **moisture movement**. Table 2.6 shows the percentage moisture movement which results when the materials are converted from dry to saturation point (maximum moisture content).

Table 2.6 Percentage moisture movement which results from dry to saturated conditions

Material	Movement per cent	
	Irreversible	Reversible
Metal, Glass	–	–
Limestone – Portland		0.004
Clay brick (typical good facing)	0.10–0.20 (expansion)	0.01
Calcium silicate bricks	0.001–0.05	0.001–0.05
Glass fibre polyester		0.02
Lightweight aggregate concretes		0.03–0.35
Dense concrete and mortars	0.02–0.08 drying (shrinkage)	0.01–0.06
Aerated concrete (autoclaved)		0.06–0.07
GRC		0.07
Timber – longitudinal		0.10
Hardboards	slight	0.11–0.32
Sandstone–Darley Dale		0.15
Perspex (ICI)		0.35
Insulating fibreboard		0.20–0.37
Laminated plastics		0.10–0.50
Aerated concrete (air-cured)		0.17–0.22
Plywoods		0.15–0.30
Softwoods: radial		0.45–2.0
tangential		0.6–2.6
Hardwoods: radial		0.5–2.5
tangential		0.8–4.0
Chipboard	large	0.1–12.0

In many cases the moisture movement is *reversible*: when the material becomes wet it expands. Upon drying, the material then shrinks back to its original size. Irreversible movement usually occurs when a material expands as it becomes wet but does *not* shrink back to its original size when it is dried again.

Self-assessment tasks 2.19

1. Referring to Table 2.6, explain *why* there is no moisture movement for certain materials, for example, metals and glass.
2. Ordinary chipboard is not used on the outside of buildings. Explain why. (The reader may wish to leave a piece of chipboard off-cut outdoors for several weeks to see what happens to it in damp conditions.)

Irreversible movement can also occur when certain materials are *dried*. Dense concrete suffers shrinkage as it dries. For this reason, it is important not to use products made from dense concrete until the drying is complete.

Observe Table 2.6 again, and look for timber. It is seen that along the direction of the grain, timber expands by 0.1% when it changes from 'dry' to saturated (this is *longitudinal* expansion). This means that the change in length is 1 mm for every 1 m of timber. However, softwoods can suffer up to 2% moisture movement in the radial direction (right angles to the grain, in a direction through the centre of a log.) In this case, up to 20 mm of expansion can take place in every metre. These are *reversible* changes, so that upon drying the material reverts back to its original size.

Before going on to the next section, the reader should browse through Table 2.6 for a few minutes to get a better feel for moisture movement.

Figure 2.16 shows the main causes of building movement. As can be seen, moisture is one of the major contributors.

Electrical properties

If a material is a good electrical conductor, it can easily allow a large electric current to pass through. A very good electrical conductor is copper. Copper wires have very low electrical resistance and are frequently used in cables. Aluminium, too, has long been in use for electrical conductors, e.g. in the grid system (see Fig. 2.17). At the other extreme, plastics (for example) are very poor conductors; they are said to be **electric insulators** and have very high electric resistances. Plastics are used to insulate conducting wires and electrical fittings inside buildings.

The ability of a given material to *resist* the flow of an electric current is called the **electrical resistance**. Resistance is measured in ohms or Ω. Good conductors have low resistances. For example, a piece of copper wire may have a resistance of just $0.000\,01\,\Omega$. On the other hand, a length of plastic may have a resistance of well in excess of $1\,000\,000\,000\,000\,000\,\Omega$ (or $10^{15}\,\Omega$). Clearly, very low resistance materials are useful for cables while very high resistance materials are used as electric insulators.

Various types of cable are used in buildings for many purposes, including the following:

- providing power for heaters, lamps, etc.
- telephone lines
- links to aerials and satellite dishes

Insulators are placed around conductors to

- protect the user from dangerously high voltage
- stop cables from touching each other, so that they do not 'short out'
- keep moisture away from the wires

Electric resistance is defined as

$$\text{electric resistance} = \text{voltage} \div \text{current}$$

So, by measuring the current which flows through a material when a voltage is applied across the two ends, the electric resistance can be obtained. In practice, in a laboratory,

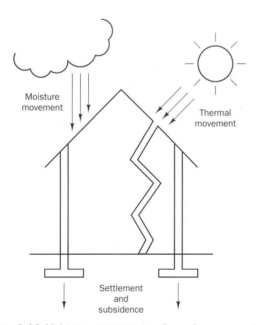

Figure 2.16 Moisture movement, thermal movement and inadequate foundations can all cause building movement. Substantial movement results in severe cracking of the structure

Figure 2.17 The cables used for transporting electric energy are made from aluminium. Can you see the ceramic insulators which are being used to support the wires?

resistance is easily measured using a multi-tester which has been set to the 'Ω' position.

Alternatively, the resistance of a material can be calculated using the formula:

$$\text{electric resistance} = \frac{\text{length} \times \text{electric resistivity}}{\text{area of cross-section}}$$

It can be seen that to calculate the resistance of a wire (say), the following must always be known:

- length (m)
- area of cross-section (m^2)
- electric resistivity (Ω m)

The last quantity is a measure of the insulation characteristics of a material. It has been measured by scientists for many types of material. Table 2.7 lists the resistivities of a number of materials. As can be seen, insulators have extremely high resistivities while conductors have very low resistivities.

Table 2.7 The electric resistivities of various materials

Material	Electrical resistivity (Ω m)
Polyethylene	10^{13}–10^{16}
Polystyrene	$> 10^{14}$
Polytetrafluoroethylene	$> 10^{16}$
Polyvinylchloride	$> 10^{14}$
Aluminium	2.7×10^{-8}
Copper	1.7×10^{-8}
Iron	9.1×10^{-8}
Silver	1.6×10^{-8}

The following are examples involving the calculation of resistance and resistivity. The reader should make sure that he or she can obtain the correct answers.

Example

A sample of a certain material has length 3 m and cross-sectional area 0.02 m^2. Taking the resistivity to be 0.1 Ω m, find the resistance across the two ends. Would this material be useful as an electrical insulator? Why?

Firstly, the formula must be written:

$$\text{electric resistance} = \frac{\text{length} \times \text{electric resistivity}}{\text{area of cross-section}}$$

Secondly, the values are substituted and the calculation is made:

$$\text{electric resistance} = 3 \times 0.1/0.02 = \mathbf{15\,\Omega}$$

Because the resistance is low, this material is *not* suitable for insulation purposes.

Example

A student was provided with a certain length of fine wire made from an unknown metal and an accurate ohmmeter (which is used to measure resistance). She made careful measurements as follows:

length = 2.17 m
diameter = 0.0002 m
resistance measured across the ends = 1.2 Ω

Calculate the electric resistivity. Hence, using Table 2.7, decide which metal was used to make this wire.

Answer: copper

Self-assessment tasks 2.20

1. Write down the units for **resistance** and **resistivity**.
2. A material X has a resistivity of 10^{17} Ω m. Which **one** of the following is correct?
   (a) X must be a metal
   (b) X can be used for power cables
   (c) X is an electric insulator
   (d) X is a good conductor

Moisture meters

Electronic instruments called **moisture meters** are used by building surveyors and other building professionals to check for damp and to measure the moisture content of materials in buildings. Figure 2.18 shows an instrument with two electrodes inserted into a timber skirting board. This type of instrument works by measuring the electric resistance of the material. Basically, as the moisture content increases the resistance decreases. For example, timber used inside buildings usually have a low moisture content. In this case the resistance is very high, many millions of ohms. Timber used outside buildings may become very wet, having high moisture content. In such cases, the resistance is much lower, typically 10 000 Ω. The electronic instruments measure the resistance and display this as a moisture content reading.

Fire resistance

Fire regulations are becoming more and more severe. Increased fire-resistance is being demanded from materials.

Figure 2.18 Many moisture meters used in the building industry work by measuring the electric resistance. In building materials such as timber, the resistance is reduced as the moisture level increases.

Figure 2.19 This fire-test house is inside one of the giant hangers in Cardington, Bedfordshire. It has been set on fire by scientists to learn more about the behaviour of materials and buildings during fires

Table 2.8 Materials classified into combustible and non-combustible groups

Combustible	Non-combustible
Timber (even if impregnated with flame retardant)	Asbestos-cement products
Fibre building boards (even if impregnated with flame retardant)	Asbestos insulation board
Cork	Gypsum plaster
Wood-wool slabs	Glass
Compressed straw slabs	Glass wool
Gypsum plasterboard (rendered combustible by the paper liner)	(containing not more than 4–5 per cent bonding agent)
Bitumen felts (including asbestos fibre-based felt)	Bricks
	Stones
Glass wool or mineral wool with combustible bonding agent or covering	Concretes
	Metals
	Vermiculite
All plastics and rubbers	Mineral wool
Wood-cement chipboards	

Source: Everett *Materials* Longman

Despite this major fires still occur. People are killed and property is damaged. Figure 2.19 shows part of the Building Research Establishment's fire research facilities where fire trials are regularly carried out in test buildings. It should be pointed out that in the event of a fire, the initial danger is usually from the combustion of the *contents* of the building. A burning settee on its own can produce as much heat as one thousand electric heaters!

Basically, fire is a **chemical** reaction. The essential ingredients needed are **fuel, oxygen** and **heat**. Many materials burn in the presence of oxygen. If more oxygen is able to reach a material, the burning takes place more quickly. Saw dust burns far more quickly than solid timber. This is because lots of oxygen molecules are able to reach the material by surrounding the small particles of wood.

Some materials, notably plastics, release poisonous gases when they burn. Victims are killed more from smoke inhalation and the poisonous gases rather than from the fire itself.

Table 2.8 should be studied. It lists both combustible and non-combustible building materials. Some examples are briefly discussed.

In Table 2.8, gypsum plasterboard, used for partition walls and ceilings, is classified as combustible because the lining for this product is made from paper. (Note that gypsum plaster on its own is classified as non-combustible.) Bitumen felt, used for flat roofing, is also combustible. Although timber is also classified as combustible, its behaviour in fire is very predictable. The speed at which timber chars is well known. This enables timber structures to be designed to resist a fire for a given time (e.g. 1 hour). Fire doors, too, are often constructed from timber because of this very predictable behaviour.

Listed in the second column of Table 2.8 are well-known examples of materials which are non-combustible. However, non-combustible does not necessarily mean that a material behaves well in a fire. It merely means that the material does not burn. Take steel as an example. Many large commercial buildings have steel frames. Although steel does not burn, it loses its strength and fails if the temperature exceeds about 550 °C. This can result in the collapse of the building structure during the fire. For this reason structural steel has to be protected by a fire-resistant, insulating wrapping. Concrete is often used to encase structural steel for this reason.

Self-assessment task 2.21

This relates to other problems which can occur where there is steel in a fire. The gaps should be filled in with the appropriate words.

In a fire, bare steel can carry large amounts of heat because it is a good thermal _____. This causes an _____ in the temperature in places well away from the fire. If material in contact with the hot steel reaches the ignition temperature, the materials then start to _____. The increase in temperature of the steel also causes significant thermal _____, resulting in bending and distortion of this material.

Bricks, and masonry in general, are very effective materials for fire resistance. These types of material are not only non-combustible but the following characteristics all help to stop the spread of fire:

- They have low thermal conductivity. This prevents the fire from spreading because the temperature of the material rises very slowly.
- The specific heat capacity (which is the heat required to raise the temperature of 1 kg of material by 1 °C) is high. This means that a lot of heat and time are required for the masonry to heat up.
- The strength of bricks is retained up to very high temperatures. In a fire, the structure remains intact. (We say that the *integrity* is maintained.) In the absence of gaps, flames cannot penetrate through the walls.

Elasticity

Whenever a force is applied, materials become stretched or compressed, or they become distorted in some other way. The more the applied force, the more they extend or compress. When the applied force is removed, many materials go back to their original size and shape. These materials are said to have **elastic** behaviour. Most materials are elastic to some extent. Steel, timber, glass, plastics, concrete, etc. all have some elastic behaviour. In fact, elastic materials behave very much like very stiff springs and they usually obey what is known as **Hooke's law**.

Scientifically, Hooke's law states that the applied force is directly proportional to the distance extended (or compressed). The graph in Fig. 2.20 is sketched for a material which obeys Hooke's law. It can be seen from Fig. 2.20 that in order to double the extension, the force must be doubled also. Likewise, in order to triple the extension, the force applied must be tripled. It is important to realise that if the deformation is elastic then the material always goes back to its original length when the force is removed.

To measure the resistance of a material to *elastic* deformation, builders, scientists and engineers use a term called the **Young's modulus**, although some people prefer the term **elastic modulus**. The Young's modulus is a measure of the stiffness for a given type of material. In order to make buildings rigid, materials with the correct level of stiffness must be chosen. Buildings which may be strong enough but wobble about are not desirable! It is important not to confuse stiffness or rigidity with strength. What is necessary for building structures is *both* high strength and high stiffness.

Table 2.9 shows for a number of relevant materials, the values of Young's modulus. As can be seen, steel is a very stiff material (it has high Young's modulus). In the previous section it was shown that steel also has high strength. High strength and high stiffness makes it a very important material for building structures. Using Table 2.9, it can be shown that mild steel is more than 12 times stiffer than lead. This is one good reason for not using lead to support the weight in building structures. (Can you think of other reasons?)

Table 2.9 Young's modulus for various materials

Material	Young's modulus (kN/mm^2)
Cast iron	250
Mild steel	200
Copper (rolled)	95
Aluminium (rolled)	62
Glass	60
Lead	16
Concrete	14 to 70
Timber	11
Rubber	0.5

The next question which needs answering is: how is the Young's modulus calculated? The formula needed is as follows:

$$\text{Young's modulus} = \frac{\text{stress}}{\text{strain}}$$

As can be seen, first the stress and the strain need to be known before the Young's modulus can be determined.

The formulae, which define the stress and strain, are as follows:

$$\text{Stress} = \frac{\text{force applied}}{\text{area}}$$

$$\text{Strain} = \frac{\text{change in length}}{\text{original length}}$$

In other words, stress is a measure of the 'concentration' of the applied force, whereas strain is a measure of the 'deformation' in relation to the original length.

Self-assessment task 2.21

From the formulae above, the units for stress and strain are respectively:

(a) N/mm and mm
(b) N/mm and no unit
(c) N/mm^2 and mm^2
(d) N/mm^2 and no unit

The reader may be thinking that to work out the Young's modulus is too tedious, but it isn't! Working through the next example will ensure that the procedure is understood. All the calculations should be checked by the reader.

Example

A piece of steel has a length of 200 mm. When a tensile force of 100 kN is applied, the length increases to 201 mm. The cross-section of this sample is square with sides measuring 10 mm. Find the stress, strain and the Young's modulus.

Answer

To find the stress:
The formula is: stress = force ÷ area
force = 100 kN
area = $10 \times 10 = 100\,mm^2$
Therefore, stress = $100 \div 100 = 1\,kN/mm^2$

To find the strain:
The formula is: strain = change in length ÷ original length
change in length = $201 - 200 = 1\,mm$
original length = 200 mm
Therefore, strain = $1 \div 200 = 0.005$

Finally, Young's modulus = stress ÷ strain
Using the values for stress and strain from above:
Young's modulus = $1 \div 0.005 = 200\,kN/mm^2$
This compares well with the value for the Young's modulus for steel in Table 2.9.

Before tackling the next self assessment task, remember that to find Young's modulus the following are needed:

- **stress** – this is calculated using
  - cross-sectional area (mm^2)
  - force applied (N or kN)
- **strain** – this is calculated using
  - original length (mm)
  - extension (mm)

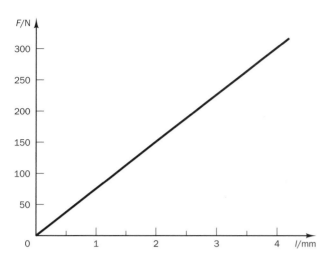

Figure 2.20 Force plotted against extension for a steel wire, diameter 1 mm, original length 1 m. The straight line from the origin indicates that Hooke's law has been obeyed

---

Self-assessment tasks 2.22

1. Find the Young's modulus, in kN/mm^2, for aluminium using the following information:

   Initial length $= 0.150$ m (length *before* force is applied)
   Final length $= 0.153$ m (length *after* force is applied)
   Force applied $= 60\,000$ N
   Area of cross-section $= 50$ mm^2
   (N.B. Assume elastic deformation.)

2. Calculate the Young's modulus for the wire in Fig. 2.20. Hence, using Table 2.9, decide which metal was used to make the wire. (Hint: to start with, find the force corresponding to any given extension.)

Although not mentioned so far, *tensile* stress and *tensile* strain are terms used when materials are tested in tension. When forces are applied in compression, the terms *compressive* stress and *compressive* strain are used.

Finally, note that certain materials such as putty and plasticine do not have elastic behaviour. If a lump of putty is pulled or squeezed, what happens after the force is removed? It remains distorted and does not go back to the original shape. This is said to be **plastic deformation**. (Do not confuse this with the materials called plastics.) When a material is undergoing plastic deformation, little or no extra force is needed to continue stretching. Bituminous materials, which are frequently used for waterproofing flat roofs, provide another example of plastic behaviour. When forces are applied to these materials (especially when they have been warmed up), they tend to 'flow'. This ensures that gaps are filled to provide water resistance.

Figure 2.21 shows the behaviour of two materials when forces are applied in tension. For mild steel, initially the deformation is elastic and then becomes plastic. Eventually, fracture occurs (it breaks). Mild steel is said to be a ductile material. For glass, only elastic deformation occurs before fracture. Glass is said to be **brittle**. Materials which are brittle can be used to support loads only when they are used in compression. If these materials are used in tension, very small cracks on the surface of the material can grow fast and this causes the material to fracture easily. Materials of this type must be **reinforced** with other materials if they are to be used in tension, or when bending forces are applied.

Self-assessment tasks 2.23

1. When forces are applied to beams, certain parts are in compression while other parts are in tension. Why must concrete beams be reinforced with steel mesh or bars?
2. Is it necessary to reinforce the concrete foundations of a house? Why?
3. Glass-reinforced plastics (GRP) are strong and tough materials, made from glass fibres embedded in synthetic resin. The plastic provides the toughness while the glass fibres provide the high strength. Why do the glass fibres not fracture if there is a sudden large impact on to a sheet of this material?

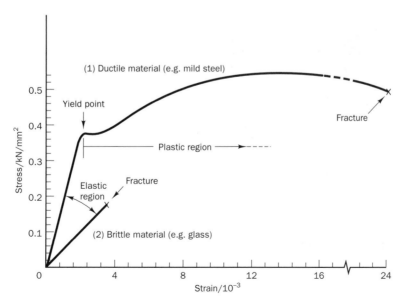

Figure 2.21 The graphs show the tensile behaviour of mild steel and glass up to the point of fracture. Glass is brittle: fracture occurs suddenly. Mild steel is tough: sudden large forces are absorbed by plastic deformation

Strength

Many forces, such as the wind and gravity act on buildings. For this reason it is essential to know the strength of materials which are subjected to large forces. Figure 2.22 shows parts of a school building that was ripped apart by strong winds. The structure was not strong enough to take the full force of the wind. The author was inside this building at the time of this disaster and he is lucky to be alive.

The force acting on a component is measured in **newtons** (N). The force of gravity can easily be found using

$$\text{force} = 10 \times \text{mass}$$

where the mass is measured in kilogram. The force of an apple pushing on your hand is about 1 N, and a single brick presses down with a force of around 30 N. The weight of an adult person pushing on the ground is typically 800 N. Remember that all these forces are caused by **gravity**, which is the most important force in building.

The **strength** of a given material is how much stress it can withstand just before it fails. More precisely:

$$\text{strength} = \frac{\text{force applied at failure}}{\text{area of cross-section}}$$

This definition includes the area (see Fig. 2.23) – this ensures that the strength does not depend on the size of

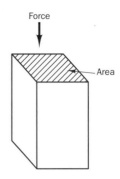

Figure 2.23 The force applied acts on the area indicated

a given type of material. For a given material, if the area is doubled, then the force required to cause failure must also be doubled – this means that the strength is the same.

From the formula for strength, it can be seen that as force is measured in newton and area is in metre squared, then strength must be measured in newton per metre squared or N/m^2. In building and engineering, the alternative unit, newton per millimetre squared or N/mm^2 is more popular.

Self-assessment task 2.23

Strength can be measured in various units including kN/mm^2 and N/m^2. Show that:

$$1\,kN/mm^2 = 1\,000\,000\,000\ (\text{or } 10^9)\,N/m^2$$

(*Hints*: 1000 N = 1 kN and 1000 mm = 1 m)

The types of forces which act on building components can be

- **compressive** or 'squashing' forces
- **tensile** or 'stretching' forces
- **bending** or 'turning' forces
- **shear** or 'sliding' forces

Figure 2.24 illustrates how the different types of force are applied. Note that when bending forces are applied to materials, some parts become stretched: these are in tension. Other parts become squashed: these are in compression. Application of, for example, bending forces can cause failure by shearing as illustrated in Fig. 2.24.

Clearly, it is important to know how building materials behave when compression, tension, bending and shear forces are applied. The measured strength depends on which particular forces are acting. For this reason, the strength of any given material needs to be known in all the different modes before being used structurally. It turns out that certain materials have high strengths in all modes. A good example is steel which is commonly used for the frames of large buildings, where all these forces are acting. On the other hand, plain concrete has high strength only in compression. It should not be surprising that concrete is used for foundations, where the forces are chiefly compressive.

Figure 2.22 Catastrophic effect of excessive forces from the wind. The remains of the roof, which consisted of a lightweight aluminium structure, can be clearly seen

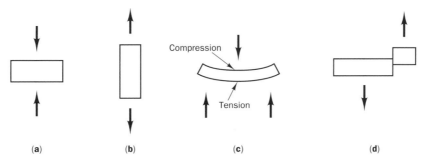

Figure 2.24 Types of forces: (a) compressive; (b) tensile; (c) bending; (d) simple failure caused by shear forces

---

Self-assessment tasks 2.24

1. Take a plastic ruler and support each end using books. This is now a simple beam, rather like the timber joists in a house. Place a small weight in the middle of the ruler to cause a bending force. Draw and label a diagram showing tensile and compressive forces on certain parts of the ruler. Indicate how the beam may fail by shearing.
2. The British Standard BS 5268 provides categories for structural timber based on the following strengths:

    1. Bending parallel to the grain
    2. Tension parallel to the grain
    3. Compression parallel to the grain
    4. Compression perpendicular to the grain
    5. Shear parallel to the grain

    Make five sketches of rectangular pieces of timber, showing the direction of the grain in each diagram. Onto each diagram, draw arrows to represent each of the forces listed above. Label the diagrams from the list given.

Many components of buildings are in compression. For example, the weight (or gravitational force) of a wall causes compressive forces on the bricks (see Fig. 2.25). It should be clear from Fig. 2.25 that bricks nearer the bottom suffer greater compressive forces than those higher up. All materials eventually fail when the compressive load exceeds a certain value. Failure may be gradual or sudden. This is further discussed later.

The **compressive strength** of a material is given by the formula

$$\text{compressive strength} = \frac{\text{compressive force at failure}}{\text{area of cross-section}}$$

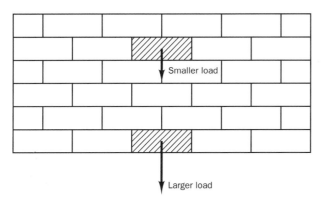

Figure 2.25 Clearly, the bricks nearer the bottom have greater compressive forces acting on them. Why?

This is just like the previous formula for strength except that the important word *compressive* has been included.

Self-assessment tasks 2.25

1. Write down two components of a house (apart from the walls) which are in compression.
2. Materials in tension often fail before the maximum force is reached. For this reason, engineers often calculate the **ultimate tensile strength** (this is the force divided by the area at **maximum load**). Write down the formula for ultimate tensile strength.

Table 2.10 shows the compressive strengths of a number of common building materials. The reader should spend a few minutes comparing the strengths of these materials.

Table 2.10 The compressive strength of a number of building materials

Name of material		Compressive strength (N/mm^2)
Mild steel		500
Plain concrete (typical mixtures)		15–40
Bricks:	common	7–14
	engineering	45–110
Timber:	softwood	30–40
	hardwood	45–60
Stone:	granite	90–140
	limestone	7–20

Referring to Table 2.10, granite has compressive strength in the range 90 to 140 N/mm^2. This means that the minimum strength expected for granites is 90 N/mm^2, while the maximum is 140 N/mm^2. Because of the variable nature of this and most other building materials, a **range** of strength is usually quoted.

Compressive failure in certain materials occurs catastrophically. Concrete suddenly cracks and can shatter. On the other hand, timber fails when the fibres begin to buckle. Figures 2.29 and 2.30 show typical failures of these two materials.

The compressive strength of materials will be discussed in more detail on page 67. However, it is necessary to try some calculations at this stage. Remember that to calculate compressive strength, both the following must be known:

- compressive force at failure (in N)
- area of cross-section (in mm^2)

Example

The compressive force on a small stone cube of sides 10 mm was gradually increased until the stone failed at a force of 2000 N. Find the compressive strength of the cube.

Answer

Compressive strength = compressive force ÷ area

compressive force = 2000 N

area of cross-section = $10 \times 10 = 100 \, mm^2$

Substituting into the formula:

compressive strength = $2000 \div 100 = 20 N/mm^2$

Example

How much force (or load) can a single engineering brick carry? Assume that this is a class B brick, compressive strength $50 \, N/mm^2$.

Answer

Assuming that the brick is laid in the normal orientation, the relevant dimensions are:

length = 215 mm
width = 102.5 mm

The formula for compressive strength must be rearranged:

compressive force = compressive strength × area

compressive strength = $50 \, N/mm^2$,

area = $102.5 \times 215 = 22\,037.5 \, mm^2$

substituting into formula:

compressive force = $50 \times 22\,037.5 = 1\,101\,875 \, N \approx 1102 \, kN$

This is equal to the force caused by roughly 110 000 kg or 110 tonnes. (Note that the symbol ≈ means 'approximately equal to'.) This is equivalent to the weight of at least ten double-decker buses!

Self-assessment tasks 2.26

1. A piece of rectangular uPVC (a very useful plastic in building) was tested in tension. The sides measured 20 mm by 20 mm and failure occurred at 22 kN. Calculate the tensile strength in N/mm^2.
2. The tensile strength of steel is about $550 \, N/mm^2$. In tension, how many times stronger is steel compared with uPVC?
3. The results of an experiment on a clay brick are as follows:

   load at failure = 280 kN

   area of smallest face = $22\,000 \, mm^2$

   What is the compressive strength of the brick?
   (a) $127.3 \, N/mm^2$
   (b) $127.3 \, kN/mm^2$
   (c) $12.7 \, kN/mm^2$
   (d) $12.7 \, N/mm^2$

When buildings are being designed, it is important to ensure that the applied forces will not approach the maximum values. Building engineers use **safety factors** which ensure that even the greatest theoretical loads stay well below the maximum load. For example, for steel, the factor of safety often used is three. Suppose that certain steel components are each capable of carrying a load up to 600 kN, as worked out from the strength. Because of the safety factor of three no component is allowed to carry a load more than 200 kN. The factor of safety frequently used for stone is 10. Why is this much greater than the safety factor for steel?

Self-assessment tasks 2.27

Table 2.11 shows 100 results for the measurement of compressive strength for a certain type of concrete.

1. Draw a **bar chart** for this data. The frequency should be on the y-axis and the strength must be along the x-axis. (N.B. The frequency is the *total number* of results at a given strength. For example, the frequency for a strength of $36 \, N/mm^2$ is 15.)
2. The main features of the bar chart should now be explained in a few sentences. (For example: What is the shape of the distribution? Is it symmetric?)
3. Use the bar chart to obtain the **median** and the **mode**. Also *estimate* the **mean**. (When a distribution is perfectly symmetric, these three averages are always equal to each other.)

Table 2.11 Compressive strengths of 100 concrete test cubes in N/mm^2

37	36	32	41	34	43	36	35	36	35
37	39	37	44	39	34	36	32	34	35
42	39	40	39	35	37	36	42	37	32
36	35	39	38	33	34	33	36	37	35
37	38	37	35	38	36	36	34	40	41
35	32	38	38	38	37	31	34	37	34
36	40	39	33	39	39	38	35	34	35
34	37	36	40	36	38	40	40	33	40
30	36	33	35	33	37	37	39	37	39
34	33	40	38	40	40	34	36	37	36

2.2 Measuring the strength of various materials

The strength of materials was defined and discussed previously. Calculations were done to find the strength of various materials. In this section, the equipment used to measure the compressive strength of materials is first described. Calculations are then done using actual results from a number of compression tests. Data is presented for timber, steel, concrete and bricks. Later, the strength is related to the actual composition and structure of these materials.

Safety is more important than anything else. The relevant safety instructions for the testing machine must be read, understood and obeyed. Ignoring the safety precautions is extremely dangerous. In particular, when materials fail, fragments may fly out at high speed. It is essential to wear safety spectacles to protect the eyes whenever compressive or tensile tests are carried out.

The tensometer

The tensometer can be used in the laboratory for the testing of small specimens of, for example, timber or steel. Figure 2.26 shows a typical tensometer.

The sample to be tested is placed and fitted as shown in Fig. 2.26. Figures 2.27 and 2.30 also show samples being tested in tensometers.

Gradual rotation of the handle causes either a tensile or compressive force to be applied. This enables an increasing compressive force to be applied. As the load increases, the mercury in the tube rises. This measures the force in kN. (Note that special mercury traps *must* be fitted to the apparatus to ensure that there is no accidental spillage of mercury.) The drum on the left turns as the sample is compressed or extended. A sheet of paper is fixed over the drum. This is used to record the results by use of a marker to prick holes onto the chart. At the end of the test, a series of holes punched onto the chart indicate the forces (Fig. 2.31 shows typical results obtained in a test).

Compression testing of timber

The test described below is for finding the compressive strength parallel to the grain of small clear samples of timber.

This type of test is very useful for comparing the strengths of different species of timber. Full details of the procedures are given in the British Standard, BS 373.

The moisture content of timber to be tested should be about 12% to 14%. The size of the rectangular test specimens should be $20 \times 20 \times 60$ mm. The samples must be cut square and accurately from the timber available. The ends of the sample must be smooth. The length must be along the grain. It is important to note the actual *measured* cross-sectional area in case it is slightly different from the nominal value (it is not good enough to accept the quoted dimensions). The species of timber being used and the direction of the applied force must also be noted. The test is then carried out as described above, under the heading The tensometer.

In such tests, results show that at first the force increases linearly, and eventually it reaches a maximum. From then onwards, the force starts to decrease; see Fig. 2.31 for typical results. (The reason for the decrease is mentioned at the end of this section.) The compressive strength is calculated by using the **maximum** compressive force applied and the area of cross-section.

As previously stated, this test is for small timber specimens which are free of knots, cracks or splits. When using timber for building purposes, the sizes of specimens are much larger so that the presence of these defects is unavoidable. The defects reduce the overall strength of the material and this must be taken into account when dealing with commercial timber.

Self-assessment task 2.28

Without looking at the notes above or below, write down the units of measurement for (a) the maximum force; (b) the area of cross-section and hence (c) compressive strength. (Check your answers when you have finished.)

Tensile testing of metals

Metals can also be tested using a tensometer. The method is similar to that described previously except that the applied force stretches the metal sample. Figure 2.27 shows a certain steel specimen during testing in a tensometer.

As in all tensometer tests, the results obtained are in the

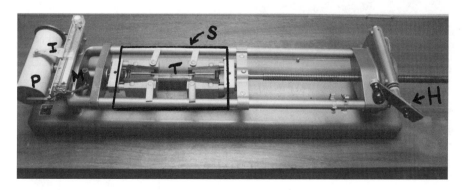

Figure 2.26 The tensometer, which can be used for testing small samples of timber in compression. Key: H; handle to apply force; T: material being tested; S: safety shield; P: paper on drum for recording results; M: mercury in tube gives applied force in kN; I: pointer used to indent the paper to record results

Figure 2.27 A steel specimen during a tensile test in a tensometer. As can be seen, the material is about to undergo catastrophic failure

form of a graph. The (ultimate) tensile strength of metals can be calculated by dividing the maximum force by the cross-sectional area.

Hydraulic testing machines

Figure 2.28 shows a large hydraulic machine used for compression testing. A machine like this can be made to apply much greater forces than a tensometer. The equipment shown can apply up to two million newtons of force. This force is roughly equivalent to the weight of twenty double decker buses! A concrete cube test sample can also be seen in position in Fig. 2.28, ready for testing. This machine can also be used

to test other building materials such as bricks. The tests for both these materials are described below.

When the compression tests are being carried out, the force must be increased steadily – but not too quickly. The compressive force applied must be increased at the rate of $15\,\text{N/mm}^2$ every minute according to the appropriate British Standard. As an example, the standard brick has a nominal face area of $215 \times 102.5 \approx 22\,000\,\text{mm}^2$. In this case, the load must be increased at the rate of $15 \times 22\,000 \approx 330\,\text{kN}$ every minute.

Compression testing of concrete

For a given mix of concrete, cubes of side either $100\,\text{mm}$ or $150\,\text{mm}$ should be produced using suitable moulds. The fresh concrete in the moulds must be compacted to remove any air gaps (voids) either by means of a vibrator or a tamping rod.

After 24 hours, the concrete cubes are removed from the mould and immersed in water for the required period of curing, usually 28 days. The sides of each cube are measured and the cross-sectional area is calculated. To follow, each cube in turn is placed into the compression machine. The load is applied as discussed above until failure occurs. The maximum load in kN is recorded for each cube. Examples of the remnants of the cubes after this destructive test are shown in Fig. 2.29.

Using the maximum load (in N) and the cross-sectional area (in mm^2), compressive strength is determined for each sample in N/mm^2. Note that the strength of concrete continues to increase for a long time after it is made. For this reason it is usual to quote the strength at a given time after manufacture of the cubes. For example, 'the strength was determined to be $37\,\text{N/mm}^2$ at 28 days'.

Full details of this test are given in BS 1881, part 116.

Compression testing of bricks

Because of the variable nature of bricks even of the same type, at least ten representative bricks should be tested. After completing the tests, both the average compressive strength and the range of strength should be calculated.

The bricks to be tested should first be numbered and then placed in a tank of water for 24 hours. The bricks are then removed and excess water is wiped off. For each brick, the length and width are measured for both the large faces and the area of each face is calculated in mm^2. If the areas of the two faces are not the same, the smaller area should be used in the calculation of strength.

Bricks should always be tested the same way as the intended use. If the bricks have frogs, these must be filled with the correct type of mortar a number of days before the test. Mortar is not required if the bricks are to be laid frog down.

In turn, each brick is mounted centrally into the compression machine, sandwiched between a pair of plywood sheets.

Figure 2.28 Hydraulic testing machine for testing bricks and concrete in compression (the safety screen was removed for the photograph)

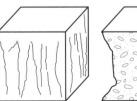

Figure 2.29 Concrete cubes after failure in compression testing

The compressive load is steadily increased as discussed above, until failure occurs. The test should not be stopped at the first signs of cracking. It is important that the load is increased until the maximum force is reached. The maximum compressive force (in kN) is then recorded. As with all these tests, the compressive strength is calculated using

$$\text{compressive strength} = \frac{\text{maximum force}}{\text{area of cross-section}}$$

The mean strength is found by adding together the strength of every brick tested and then dividing by the number of bricks.

As previously stressed, safety is most important. Safety spectacles are essential and a safety screen must be present throughout the tests. Emergency stop buttons should be available to cut the power in the case of an emergency. Regarding the testing machine, the manufacturer's instructions must *always* be strictly followed.

Full details of this test are given in BS 3921.

Results of various strength tests

Results from various compression tests are now presented:

Steel

The following are results from a tensometer test on a certain steel specimen, diameter 4.5 mm. The (ultimate) tensile strength can be calculated as follows:

$$\text{maximum force} = 11 \, \text{kN}$$
$$\text{area of cross-section} = \pi \times \text{radius}^2 = \pi \times (4.5 \div 2)^2$$
$$= \pi \times (2.25)^2$$
$$= 15.90 \, \text{mm}^2$$

So,

$$\text{strength} = \text{maximum force} \div \text{area} = 11 \div 15.90$$
$$= 692 \, \text{N/mm}^2$$

Self-assessment task 2.29

What is the ultimate tensile strength (stress) of the mild steel in Fig. 2.21. Ensure that you record the answer using the appropriate unit. Note that calculations are *not* necessary here.

Timber

Figure 2.30 shows a piece of timber after testing using a tensometer. It can be clearly seen that failure occurs by the buckling of the fibres. The load that the timber was able to carry then started to fall. Figure 2.31 shows the data obtained for a test sample. The *y*-axis represents the force in kN and the *x*-axis shows the reduction in length as the force was applied. (The reduction in length is highly magnified.)

Using information from Fig. 2.31, the compressive strength can now be calculated:

Force at failure (from the graph) = 16 kN = 16 000 N
Area of cross-section = 20 × 20 = 400 mm²

Figure 2.30 Timber specimen after failure in compression test. Note that the force was applied in the direction of the grain. The buckling of the fibres can be clearly seen

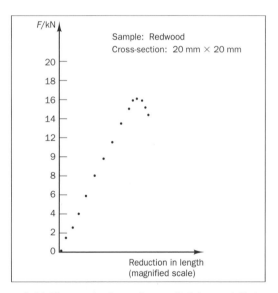

Figure 2.31 The graph shows the applied force plotted against the reduction in length. The sample used was redwood, free from knots and other defects

So,

$$\text{compressive strength} = \frac{\text{load at failure}}{\text{area of cross-section}}$$

$$= \frac{16\,000}{400} = 40 \, \text{N/mm}^2$$

Self-assessment task 2.30

The test is to be carried out several more times with other similar samples. Do you think that the results would give exactly the same value for the strength? Why?

Clay bricks

Table 2.12 shows the results of tests on individual bricks of several different types. As can be seen, the length and width of both faces have been measured in each case (see Fig. 2.32 for the definitions). All calculations in Table 2.12 should be checked by the reader. As discussed previously, ten bricks of each type should normally be tested. The average compressive strength must then be calculated from the results.

Table 2.12 Results of compression testing of several types of brick

Brick type	Dimensions (mm)				Smallest area (mm²)	Max. load (kN)	Strength (N/mm²)
	l_1	w_1	l_2	w_2			
Fletton common	216	103	215	103	22145	470	21
Engineering (A type)	214	102	214	103	21828	1510	69
Engineering (B type)	215	103	215	103	22145	1150	52
Dorking stock	217	102	216	102		1100	

The blanks should be completed by the student. See the self-assessment task below.

Self-assessment tasks 2.31

1. In Table 2.12, the calculations for Dorking stock have not been completed. Work out the smallest area and the strength.
2. Ten representative bricks of a certain type were tested. The calculated values of the strengths in N/mm² are as follows: 35.1, 36.3, 35.5, 38.6, 36.4, 37.4, 36.8, 36.5, 34.9, 35.5
Determine the mean compressive strength.

Concrete

Before looking at some data from concrete testing, the following self-assessment task should be tried. The student should remember the answers from earlier reading.

Self-assessment task 2.32

Fill in the blanks:
Concrete test blocks are usually _____ in shape and have sides either _____ mm or _____ mm. These are produced using special _____. In preparation for the tests, the freshly made concrete is placed into the moulds and compacted to eliminate _____. After _____ hours the concrete cubes are removed from the moulds and cured by immersing in _____, usually for up to 28 days after manufacture. The compression tests are often carried out at _____ days. These test procedures are standardised so that results from different tests can be compared with one another.

Specimen results

Figure 2.29 shows the remnants after typical (correct) failures of dense concrete blocks. Test data for a particular block are as follows:

Age: 28 days
Size: 150 mm cube
Load at failure $= 110$ tonnes $= 110 \times 10\,000 = 1\,100\,000$ N
(N.B. 1 tonne is equivalent to 10 000 N.)

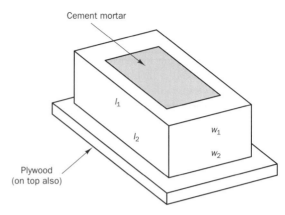

Figure 2.32 Brick before compression test. For the calculation of compressive strength, the smaller one of $(l_1 \times w_1)$ or $(l_2 \times w_2)$ must be used

$$\text{Compressive strength} = \frac{\text{load at failure}}{\text{area}}$$

$$= \frac{1\,100\,000}{150 \times 150} \approx 49\,\text{N/mm}^2$$

Self-assessment tasks 2.33

Ten concrete cubes (sides 150 mm) were destructively tested to determine the compressive strength. The loads, in tonnes, for failure were as follows:

97 112 104 110 98 102 100 105 106 106

1. Find the mean (average) compressive force in kilonewtons at failure. (Remember: 1 tonne is equivalent to a load of 10 kN.)
2. Find the mean compressive strength in N/mm².

The following data (Table 2.13) compares dense concrete made using gravel aggregate with a lightweight concrete made from foamed blast furnace slag. Figure 2.33 shows the appearance of a piece of the lightweight concrete.

Table 2.13 Comparison of two types of concrete

	Bulk density (kg/m³)	Compressive strength (N/mm²)	Young's modulus (N/mm²)
Dense concrete	2200–2500	14.0–70.0	21 000–35 000
Lightweight concrete	1000–1500	2.0–7.0	6 900–21 000

The data shows that the range of compressive strengths for the dense concrete is roughly ten times that for the lightweight concrete. Try to think why this should be the case – the reasons are explained later.

Self-assessment tasks 2.34

1. Compare the stiffness and density of the two types of concrete in Table 2.13.
2. The lightweight concrete in Table 2.13 is to be used for the construction of load bearing walls by casting in-situ. A certain wall measures 305 mm deep and is 5 m long. Calculate the maximum load which can be carried assuming the minimum compressive strength for the concrete.
3. Give one major advantage in using lightweight concrete for the construction of walls.

Figure 2.33 Lightweight concrete made from foamed blast-furnace slag

How does the composition and structure affect the strength?

Metals

Metals are **crystalline**. This means that they are made of crystals. Usually the crystals are too small to be seen by the naked eye. However, large zinc crystals are often visible on the surface of galvanised steel (e.g. on fencing, buckets, etc.). Figure 2.34 shows zinc crystals. Smaller crystals of metal can sometimes be seen on the surface of polished brass door knobs.

Crystals form when atoms are arranged regularly, or in

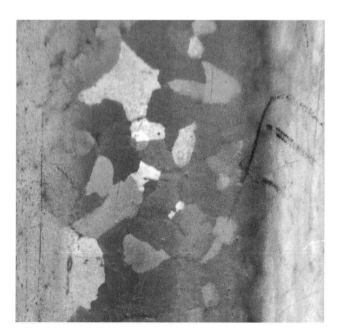

Figure 2.34 Large zinc crystals are clearly visible on the surface of this galvanised steel section. Readers should try to make their own observations of metal crystals

patterns. This is the case in metals. Figure 2.35 shows a simple arrangement for the atoms in a metal.

The arrangement shown in Fig. 2.35 can explain the *elastic* behaviour of metals (see page 62). As a tensile force is applied, the distance between adjacent atoms increases slightly. When the force is removed, the atoms return to their normal positions.

If metal crystals were 'perfect', as shown in Fig. 2.35, then plastic deformation (giving ductility to metals) requires entire layers of atoms to slide over each other. Scientists have shown that much more energy than is available is required for this to happen. So, this is *not* the correct explanation for plastic deformation. It is rather like trying to move an entire carpet across a room by pulling at one end!

In fact, plastic deformation results from *defects* in crystals. Figure 2.36(a) shows a 'defective' crystal, where the upper portion contains an 'extra' plane of atoms. When forces are applied as shown, the dislocation shown moves along to the right, as can be seen in Fig. 2.36(b) and (c). This effect results in the permanent deformation of the material. This explanation is quite acceptable from the energy point of view. It is the presence of dislocations within the metal crystal, then, that causes plastic deformation. Using the carpet analogy again, this is like introducing a kink into the material. Pushing the kink along is quite easy and the carpet is displaced once the kink has reached the other end.

The presence of dislocations in metal crystals means that *pure* metals are generally not strong enough to be used structurally. Metals are made stronger and stiffer by introducing 'alien' atoms into their crystals. The presence of impurities obstructs the movement of dislocations, making the material much stiffer and stronger. A small, controlled, amount of carbon is introduced into iron for this reason. The result is **steel**, a very significant building material. The percentage of carbon can be varied to alter the properties of the steel. As another example, pure aluminium has low strength. However, various alloys are produced (e.g. Duralumin) by adding impurities such as manganese and copper. This makes the material much stronger. Aluminium alloys are important engineering materials.

Timber

Figure 2.37 shows a sketch of a softwood log. The transverse, radial and tangential faces can be clearly seen. Figure 2.38 shows the microstructure of both a softwood and a hardwood. The reader should be able to relate the faces to

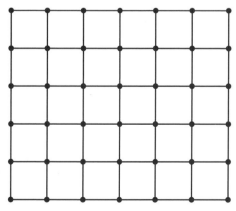

Figure 2.35 Atoms forming a single perfect crystal. The atoms are held in place by interatomic forces

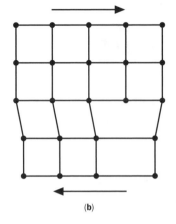
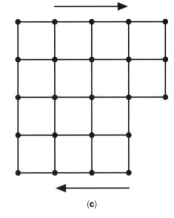

Figure 2.36 Plastic behaviour is explained by the motion of dislocations through a metal crystal

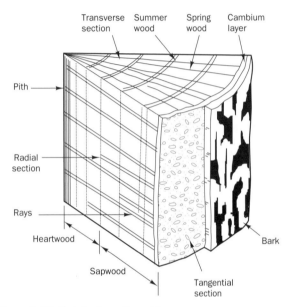

Figure 2.37 Sketch of a softwood log

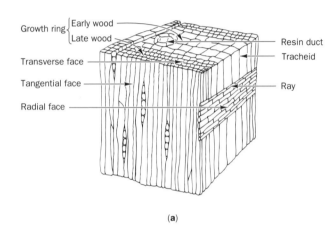

(a)

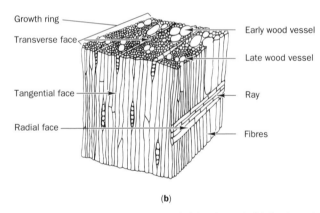

(b)

Figure 2.38 Microscopic structure of: (a) softwood; (b) hardwood

those shown in Fig. 2.37. From the diagrams, it should be clear that timber is an **anisotropic** material. This means that its behaviour depends on direction (more on this below).

In timber, the fibrous material (which is mainly cellulose) runs along the direction of the grain and the material is held together by lignin, which helps to stiffen the structure. In softwoods, the vertical elements are called tracheids and are arranged fairly regularly. In hardwoods, the fibres have a variety of sizes.

Clearly, the strength of timber depends on the direction in which the force is applied (timber is anisotropic). The compressive strength when the force is applied parallel to the direction of the grain is considerably greater than when the force is applied perpendicular to the grain. (Typically, the compressive strength is about three times greater in the direction of the grain compared with the strength at right angles.)

The reason is that the long cellulose molecules are aligned mainly along the grain direction. In this direction, the forces holding the molecules together are very strong. This is known as **covalent bonding**. The strength at right angles is much smaller because the forces between the molecules are much smaller. This is called **hydrogen bonding**.

The reader may wish to try the following test in order to understand the behaviour of timber. A bundle of, say, twenty straws is required. Place a rubber band at each end in order to hold the 'tracheids' together. It is easy to get a feel for the strengths by squeezing the material both along the 'grain' and at right angles.

The strength along the length is much greater than that at right angles to it.

As an illustration, Douglas fir, which is an important timber used in building, has compressive strengths for large samples as follows:

strength parallel to the grain $= 10.5\,\mathrm{N/mm^2}$
strength perpendicular to the grain $= 2.4\,\mathrm{N/mm^2}$

Timber products: plywood and chipboard

Plywood is a composite material made from timber and adhesives. Veneers, which are thin sheets of timber, are glued together to produce the plywood. If the plywood is made using resin adhesives, the material does not deteriorate in the presence of moisture and can be used externally (e.g. soffit boards). Timbers frequently used to produce plywoods include Douglas fir, gaboon, beech and birch.

Figure 2.39 shows the structure of plywood. Notice that there is an odd number of layers. This means that the material is symmetric about the centre, ensuring a 'balanced construction'. As a result of this, moisture movement in plywood is minimised. In solid timber, however, moisture movement can result in serious distortions, including warping.

The direction of the grain of each ply is at right angles to that on adjacent plies. This feature of plywood ensures that the strength is similar along both the length and width of the board. Compare this feature with solid timber.

Many other types of timber-based materials exist. These include **particle boards** and **fibre boards**. Unlike solid timber, the strength properties along both the length and width of these boards are the same. What does this imply about the structure of these materials? The composition of particle boards and fibre boards is discussed briefly on pages 81 and 82.

Bricks

Bricks are manufactured mainly from clay (which consists of silica and alumina), where water and other chemicals may be

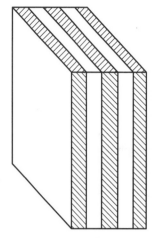

Figure 2.39 The structure of plywood (five ply): the laminations are held together with adhesives

added. The raw material is ground and then pressed into moulds.

Certain types of bricks are produced by an extrusion process. In this case the raw material leaves the machine in a continuous strip. Stretched wires are used to cut up the material into the size of individual bricks. The resulting bricks are then fired at high temperatures, between about 800 °C and 1300 °C. Various chemical reactions take place during the firing process. At the end of the heating period, once the fired bricks have cooled down, they are ready for use.

Some bricks are solid. Other types have a frog, i.e. a depression in one face. Extruded bricks often have holes passing through them; see Fig. 2.40. (Holes can take up to about 25% of the overall volume.) The presence of frogs or holes reduces the amount of material needed per brick and makes them easier to handle. Although holes can reduce the strength of bricks, the porosity affects the strength considerably more.

Bricks are porous materials. This means that they contain lots of pores or tiny holes (or voids). Porosity was previously discussed on page 57.

Pores in bricks and other ceramic materials are areas where large stresses build up when forces are applied. When the stress at a pore reaches a critical value, a crack forms and grows until the material fractures. In addition to the pores, the presence of minute surface cracks also greatly affects the strength.

Because of the porous nature of bricks, water can be absorbed by **capillary action**. As more water is absorbed, the compressive strength of bricks is reduced. Low-porosity, high-density bricks tend to have high compressive strength. Also, bricks with low porosity usually have high durability.

Engineering bricks are an excellent example of low

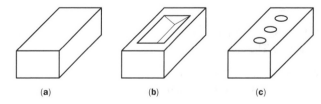

Figure 2.40 Several kinds of brick: (a) solid; (b) with frog; (c) with holes

porosity, high density bricks. These are used where high compressive strengths and/or high durability are demanded.

Concrete

Unlike materials such as timber, this material has isotropic behaviour, as it appears uniform from all directions. This means that the strength does not depend on the direction of the applied force.

Concrete is a brittle material. Plain concrete is therefore used in compression and not in tension.

Plain concrete is produced from a uniform mixture of water, cement and aggregates. Upon setting and hardening, the aggregate particles become held in place by the cement. The cement is initially in powder form and is responsible for binding together the entire mixture. The cement is the 'active' ingredient in the mixture. When it is mixed with water a chemical reaction starts. The fresh concrete begins to set and then harden. The strength is developed gradually and the maximum strength may not be reached for many months.

There are many types of cement. Portland cements are the most frequently used. 'Ordinary' Portland cement is a low cost, general purpose cement. It is widely used in building.

The aggregate often consists of small stones and sand. The stones are of varying sizes. Having a range of sizes ensures that the gaps produced by the packing together of larger particles are filled in by the smaller particles. The main reasons for the use of aggregates in concrete are as follows:

- they reduce the overall cost of materials as aggregates are considerably cheaper than cement
- aggregates reduce thermal movement and also shrinkage during drying
- aggregates improve the resistance to wear (durability) by both abrasion and weathering
- the appearance of the concrete can be altered by using different types of aggregate

The next assessment task should be completed to ensure that the reasons for the ingredients in concrete are understood.

Self-assessment task 2.37

Fill in the missing words:
Concrete is produced by the use of cement, water and _____ . Cement is the material that _____ the concrete together. Water is needed to start the _____ reaction which eventually results in the hardened concrete. Aggregates usually consist of _____ and sand. The inclusion of aggregates is important for various reasons. In addition to reducing cost, aggregates reduce drying _____ and improve the durability of the concrete.

What factors determine the compressive strength of the concrete? The answer is

- the type of aggregate used
- the free water/cement ratio

The second reason is more important. The *free water/cement ratio* is just the mass of water added to the concrete mixture

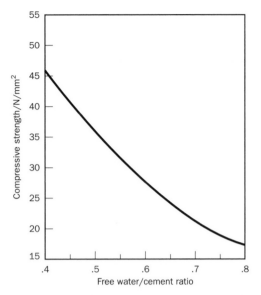

Figure 2.41 Compressive strength plotted against free water/cement ratio for test cubes at 28 days

divided by the mass of cement used. As an example, let the mass of water in a certain mixture be 40 kg and the mass of cement 80 kg. The free water/cement ratio is $40/80 = 0.5$.

The smaller the value of the free water/cement ratio, the greater becomes the strength of the concrete. Figure 2.41 shows this clearly. The problem is that if the water content is too low, it becomes very difficult to compact the fresh concrete – this means that some of the air gaps remain. If the fresh concrete is not fully compacted, these air voids reduce the final strength. For this reason there must be sufficient water to ensure that the *workability* of the fresh mixture is adequate. Figure 2.41 assumes that the material has been fully compacted.

At the other extreme, if the free water/cement ratio is too high, air voids are formed when the excess water eventually dries out; the material is now more porous. This effect clearly reduces the strength of the concrete. So, all other factors being the same, a mixture with a free water/cement ratio of 0.5 will produce stronger concrete than one with ratio 0.8. Figure 2.42 shows how the strength of concrete is reduced as the air voids increase. Concrete which contains just 5% by

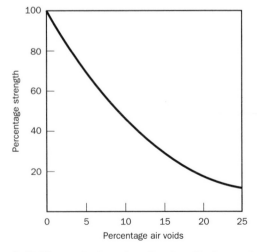

Figure 2.42 The graph shows how the strength of concrete varies as the percentage of air voids increases

volume of air has about 70% of the strength of the same concrete which contains no voids.

The presence of pores reduces the strength because they create high stress regions. Cracks are produced when the stresses become very large. The cracks grow until the material fractures. This effect was also discussed above, under the heading 'Bricks'.

In summary, the lower the free water/cement ratio the greater the strength of the concrete. However, if the free water/cement ratio is too low, the material cannot be fully compacted and air voids remain. The voids always tend to reduce the compressive strength.

The next self-assessment task will check your understanding of the importance of the free water/cement ratio.

Self-assessment tasks 2.38

1. Explain briefly what the graph in Fig. 2.41 shows.
2. By approximately what percentage does the strength of concrete *decrease* when the percentage of air voids is increased from 5% to 15%? (Use Fig. 2.42 to obtain the answer.)
   (a) 70%
   (b) 60%
   (c) 50%
   (d) 40%
   (e) 30%
3. Although reducing the water/cement ratio increases the strength of concrete, why should very low values of the water/cement ratio not be used?

2.3 Selecting materials for building elements

Introduction

First in this section, the important components of buildings are introduced. Particular attention is given to the elements of low-rise dwelling homes. Low-rise domestic buildings are usually one-, two- or three-storey houses which may be detached, semi-detached or terraced.

Later, the discussion focuses on the materials used, and the reasons for choosing them. It is important that the reader doesn't just try to remember the names of materials and where they are used. The student should also refer to the material properties described earlier in this Unit. This will help in understanding the reasons *why* certain materials are selected for a particular use.

Building elements

The important parts of a building are called the **building elements**. The building elements of houses include:

- foundations
- external walls
- pitched roof
- windows
- partition walls
- suspended floors
- services

As can be seen, many of the elements face the external environment. These external components have essential uses. Figure 2.43 shows some of the purposes of these components.

The various building elements are now briefly discussed in turn.

Foundations The foundations are below ground level and support the weight of the building. Foundations are wider than the walls which are supported. This reduces compressive stresses to a level that weaker subsoils can bear. If correct foundations are not provided, the structure will gradually push the soil down and subsidence will occur: this causes walls to crack and distort (see Fig. 2.16).

External walls These have a number of purposes:

- to provide a rigid structure for the building and to support the roof and the floors
- to cut out the wind and be able to resist wind loads
- to stop water penetrating into the house
- to insulate against heat loss from the inside of the house
- to stop loud noises entering (or leaving) the building

Pitched roof The roof stops the penetration of rain and snow. In order to efficiently drain away the rain water, the angle of the roof must be more than about 35° measured from the horizontal. The roof must be strong enough to resist the large loads produced by the wind and snow.

Windows The windows allow daylight to enter the inside of the house. Windows which can be opened permit ventilation when required. Also, the windows allow the occupants to view the outside environment.

Partition walls The inside of a house is usually divided into rooms by the use of partition walls. This allows for privacy and also reduces noise and smells penetrating other parts of the house. Depending on the structure, certain internal walls may also be load bearing.

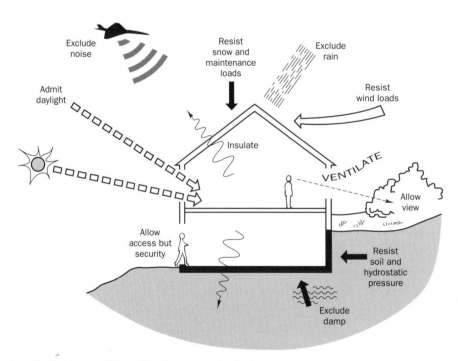

Figure 2.43 Some important purposes of the external components of a house

Suspended floors The floors provide support for the loads produced by people, equipment, furniture, etc.

Services People in houses need energy, water, sanitation and means of communication with the outside world. These are known as **services**.

Energy is delivered to homes by:

- underground cables carrying electricity
- underground pipes delivering gas

Mains water arrives in underground pipes. The majority of homes in the UK have an indirect cold-water distribution system. This means that the mains water is fed directly to certain taps, for example, the cold water tap in the kitchen. The rest of the water is diverted to a large cistern in the loft. The cistern feeds the hot water cylinder, toilets, bath taps, shower, etc.

The meaning of **sanitation** is to drain away and dispose of sewage, i.e. waste water and waste solids. This is achieved by use of various pipes from, for example, wash basins and toilets. These lead to a vertical large bore pipe (soil pipe), which runs to the external sewer.

Telephone lines are used for communications. Low-power cables, and increasingly, optical (glass) fibres are used for transporting the electronic signals.

The reasons for selection

Clearly, there is a range of materials needed for the various building elements. How are the materials chosen? The important reasons for choosing a given material are:

Performance in use

Each material used must have the correct properties for the intended use. This is why the properties were extensively described and discussed in sections 2.1 and 2.2. For example, external, load supporting walls must be strong and rigid. This means that stiff (rigid) materials must be used which have high compressive strength. Needless to say, these properties must be retained over the lifetime of the house. As another example, the roof is designed to stop water penetrating into the house. The materials used must be capable of resisting water for many, many years. Loft insulation is used to reduce the heat escaping from the roof of a house. The important property here is thermal insulation. The material to be used must have very low thermal conductivity. Also, this property must be retained for a very long time. Summing up, materials must not only have the suitable properties needed for the job – they must be durable. These properties must be retained over (hopefully) the lifetime of the house.

Cost

Economics, or the cost of producing a building, is always of primary consideration. No building company will start a project unless a profit is going to be made! In producing domestic buildings about half the cost goes on materials and the rest is the labour cost. (In commercial and industrial buildings the cost of materials is considerably greater than the labour cost.) For this reason, the materials chosen must not only have the necessary properties – they must be affordable.

It should be noted again that domestic buildings are designed to last for a very long time. If poor quality, low-durability materials are used during construction, the overall *long-term* cost of materials becomes high. This is because the materials quickly degrade or decay. It then becomes necessary to replace the materials or to carry out repairs. For this reason, it is important to consider the *long-term* maintenance costs as well as the *initial* building costs when choosing the materials.

Availability

Materials should be selected for building not just because they have the right physical properties. The **availability** must also be questioned.

It is well known that Eskimos build their houses using snow. To them, this is one of the few materials available and fortunately it is very effective in protecting them against the weather. Throughout this unit we have seen that in this country we have a wide range of building materials available for our use. However, availability is still an issue. Here are a couple of examples. Certain types of timber which were widely used in the past are now hardly available. Brazilian mahogany was used extensively in the past. Today, there is little of this particular resource remaining. Its use has greatly declined because exploitation has resulted in a dramatic reduction in availability.

Many of the buildings in the historic city of Bath are built from varieties of limestone. Why? It should be of no surprise that limestone is locally produced in that region. The ease of availability has 'forced' this city to be built largely using this material. Aberdeen is sometimes called 'the granite city'. Referring to Fig. 2.44, where do *you* think the cities of Bath and Aberdeen may be situated? Now refer to a road map to find out whether your 'guesses' are correct.

Self-assessment task 2.39

Traditional housing in England has employed bricks, blocks and mortar. In North America, the majority of homes are built from timber. Give possible reasons for this.

Environmental considerations

In recent years, people have become increasingly aware of environmental problems. Some of the current issues are:

Global warming The burning of fossil fuels produces an increased amount of certain gases such as CO_2. This leads to the greenhouse effect resulting in temperature increases in the environment. With the sea level apparently rising, this could have a devastating effect on the environment in the long term.

Ozone depletion Use of certain industrial gases, for example CFCs, is resulting in the reduction in the thickness of the ozone layer in the earth's upper atmosphere. This means that more ultra-violet radiation from the sun is able to reach the

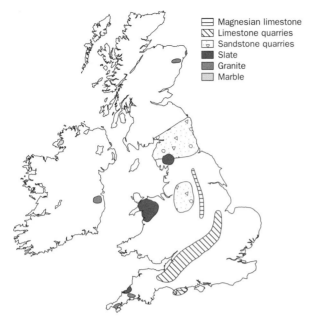

Magnesian limestone	
Limestone quarries	
Sandstone quarries	
Slate	
Granite	
Marble	

Figure 2.44 Major quarrying regions of building stone in the United Kingdom

earth's surface. As a result, problems such as skin cancer are said to be on the increase. CFC gases were previously used in the production of foamed plastics. (These materials are used by the building industry as thermal insulation.) Fortunately, industry has now largely switched over to using 'ozone friendly' gases.

Diminishing resources The earth's materials are not limitless. Over-use and misuse of certain materials means that certain non-renewable materials (e.g. oil, coal) are declining rapidly. Alternative materials must be sought.

Processed or manufactured materials (e.g. cement, metals, bricks, glass, plastics, etc.) require large amounts of energy for their production (and transportation). Some processes require more energy than others for the manufacture of a material. Take aluminium, which is the most abundant metal in the earth's crust. Huge amounts of electrical energy are required for its purification from the mineral state. (Use of electric power means more fossil fuels are consumed and more greenhouse gases released into the atmosphere.) Consequently, aluminium is far more expensive than iron and steel.

Tropical rain forests are at present being cut down at an alarming rate. This may provide long-term environmental problems. However, timber is one of our few renewable resources. With properly managed forests timber supplies can last for ever. Timber is an 'environmentally friendly material'.

In summary, when choosing materials for building, it is important to consider the impact on the environment. Careful selection will ensure that the environment remains in good shape for many future generations.

Appearance

It is important for buildings to look good. Architects spend their lives designing buildings to look attractive and to blend in with their surroundings. Also, they make sure that the external appearance matches the purpose of the building.

The appearance of a building comes from its structural form, that is its size and shape and the way it is put together. The appearance also depends very much on the texture and colour of the surfaces of the materials used. The 'looks' of a building are, however, quite subjective: what is attractive to one person may be ugly to another. The appearance of some large modern buildings has caused much controversy.

At the present time, most low-rise domestic buildings are still built with traditional materials, for example, bricks and timber. These materials have 'natural textures', and because they look familiar, people are generally happy with the external appearance of houses. Figure 2.45 shows a traditional design domestic property built with traditional materials. Figure 2.46 shows non-traditional 'cubic houses'. These houses are built with many non-traditional materials. They appear quite different to traditional houses. Which type do you prefer? Why?

Figure 2.45 A traditionally built modern house in England

Figure 2.46 Non-traditional housing in Holland

Conditions of use

The various elements of a domestic building (listed on page 76) are used under a variety of conditions. Certain elements are load bearing; as we have seen the external walls need to support the weight of the building against the force of gravity.

Various elements are exposed to the harshness of the weather and they must therefore be able to resist the problems this causes throughout the lifetime of the building. Exposure to weather can mean various things. We have seen previously that external building elements as a whole must be strong enough to overcome the force of the wind. The weather also causes changes in temperature: the sun in the summer can heat certain parts of the building to very high temperatures (it can get hot enough to fry eggs on the roof tiles!). On the other hand, cold winds and the lack of sun in the winter cause temperatures to fall to very low values. As a result of these temperature changes substantial thermal movement occurs in some elements. In addition, the weather frequently provides rain. The presence of water can cause moisture movement, damp problems and frost damage.

The external elements are frequently exposed to chemicals in the atmosphere. The performance of the materials chosen must not be substantially affected by the chemicals. Those elements exposed to the soil below ground level must not be damaged by the chemicals present in the soil.

Various elements of the building must have a specified resistance to fire. This is expressed as the time during which an element can resist a fire. The fire resistance of an element is measured by the following:

- stability: the resistance to collapse
- integrity: the ability to resist the penetration of flames
- insulation: this is the resistance to a rise in temperature

The elements which must be fire resistant are outer walls, partition walls and floors. The Building Regulations stipulate a 30 minute fire resistance for these elements. Note that the load bearing elements must be able to resist collapse while carrying maximum load throughout the heating period. Components which are glazed (e.g. windows and some doors) are classed as having no insulation against fire. The fire resistance offered by the roof depends on how close other buildings are situated. The fire resistance should provide protection against the spread of flames either out of or into the building. In the case of a fire, the roof should be able to withstand early collapse.

Self-assessment task 2.40

Make a list of the building elements which have been listed above and write the relevant phrases for the conditions under which they are used. The phrases are: load bearing (or non-load bearing), exposed to the weather, fire resistant.

Examples:
External walls: load bearing, exposed to the weather, fire resistant
Partition walls: (mainly) non-load bearing, fire resistant

In the next section, the various building materials are discussed together with their uses for the building elements in houses. Readers should study this section carefully,

referring to sections 2.1 and 2.2 where necessary. By the end, it should be clear exactly why the chosen materials are used. The self-assessment tasks will check the student's understanding.

Materials used for the building elements

In the final part of this unit, the materials required for the various elements in domestic buildings are first summarised. This is designed to give the reader an overall feel of how the various elements 'fit together'. Materials are then discussed under these seven major headings:

- timber
- timber products
- clay products
- concrete and concrete products
- metals (ferrous and non-ferrous)
- bituminous materials
- plastics (thermoplastics and thermosetting plastics)

Figure 2.47 shows a cut-away view of a typical domestic property. There are, of course, many variations from the type shown.

Starting at the bottom, the strip foundations are usually made from non-reinforced concrete. (Can you remember the purpose of the foundations? If not, read page 76 again.)

The ground floor is made from concrete slab (around 100 mm deep) which usually sits on hardcore (broken stone). For large areas the concrete slab is often reinforced using a steel mesh which provides stability. In older type properties, the ground level is usually a suspended timber floor (see Fig. 2.47).

The walls have an inner leaf and an outer leaf, with an air cavity between them. These are known as cavity walls. The outer leaf is made from bricks and mortar. The inner leaf is composed from lightweight blocks (clay or concrete). Metal or plastic ties are used at about 1 m intervals to fasten the inner and outer leaves to improve the stability of the structure. To reduce heat losses to acceptable values, insulating material (e.g. expanded polystyrene) is required in the cavities. A damp-proof course (e.g. polyvinyl chloride sheet) is inserted between two layers of bricks. This is near to, but above, ground level. The damp-proof course prevents moisture moving up the brick wall, and stops rising damp.

The first floor is supported on timber joists which themselves are supported by the load-bearing walls. In the past, floors were made from tongue-and-groove floor boards. Flooring grade chipboard has replaced timber floor boards in recent years.

Glass (4 mm thick) is almost always used in the windows but the frames can be made from a number of materials such as timber (both softwood and hardwood frames exist). Metal frames can be constructed from galvanised steel (this is steel which is coated with a thin layer of zinc to stop rusting) or lightweight aluminium. Aluminium frames are normally mounted within hardwood surrounds.

The roof is pitched and consists of triangular trusses made from timber. The timber frames are fixed to the walls by metal straps. Tiles (made from concrete or clay) are fixed onto the roof by nailing them into battens. They are staggered

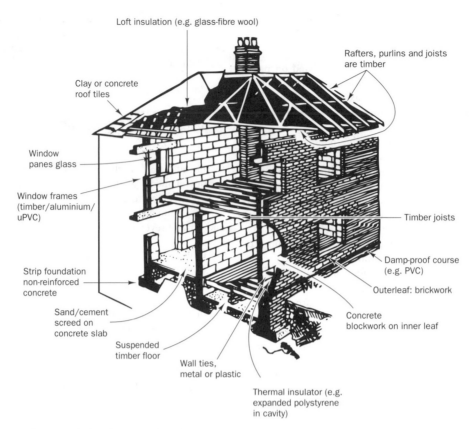

Figure 2.47 Cut-away diagram of a traditional brick-joisted type of domestic property

Figure 2.48 Loft insulated to reduce heat losses

and are overlapping to minimise rain penetration. Bituminous felt is present below the tiles as further defence against water. Blankets of glass-fibre wool (e.g. 150 mm thick) are placed into the loft for insulation purposes. Figure 2.48 shows a typical insulated loft area.

Within the structure, the house is divided into rooms by means of partition walls. Some partition walls are load bearing but the majority are non-load bearing. Partition walls may be constructed from concrete blocks or hollow clay blocks. Concrete blocks used for partitions are normally lightweight or low density. (The reader should be able to think of at least three advantages in using *low density* blocks for partitions.) Sometimes bricks are used although their use nowadays is normally limited to load-

bearing partitions and where a high level of fire resistance is needed. Timber stud partitions are an alternative to the types already mentioned. A timber stud partition is simply a lightweight frame made from timber. Either hardboard, plywood or plasterboard is fixed to the frame to form the surfaces of the partition.

Finally, a brief look at the materials for services. For the water supply, a service pipe brings water to the house. The service pipe can be made from copper, galvanised steel or plastic (uPVC or polythene). Metal pipes have to be protected against corrosion, by for example, encasing them inside a plastic covering. Soil pipes, which carry away waste water and waste solids, are made from galvanised steel, cast iron or uPVC. Service pipes for the delivery of gas are normally made from galvanised steel. Inside the house, the pipework is either mild steel or copper. Electricity arrives by means of underground cables. Copper is the usual material used to carry the electric current. (Copper is an excellent conductor of electricity.) The copper conductors are contained inside plastic materials which are excellent electrical insulators. Likewise, the cables inside the house use copper with PVC as the insulating material.

The main groups of building materials, their uses, and the reasons why they are selected are now discussed in more detail.

Timber

At various points in this unit it has been shown that timber has a number of advantages over many other materials. Some of these are:

- timber is a natural, 'environmentally friendly', renewable and easily available material
- timber is lighter than most other materials
- the 'strength to weight ratio' is high both in tension and compression
- this material is easy to cut and shape
- although timber burns, its behaviour in a fire is very predictable
- timber is a good thermal insulator – hence it has a warm feel

Suspended (upper) floor The density of timber is very low (its density is only about 7% that of steel). The strength in bending is substantial. Timber is a fairly stiff material and it behaves elastically. Also, being very easy to cut and shape, this makes timber a very suitable material for use in suspended floors.

The timber joists are the beams which carry the weight of the floor boards, furniture, equipment and people. (The reader may wish to lift up a floor board to view the joists. Note that permission may be needed!) The timber joists are themselves fixed onto or into the load bearing walls which support the suspended floor. Joist sizes vary, but can be up to 75 by 225 mm in cross-section. Joists are fitted 'deep', that is the wider sides of the joists are vertical. This gives the material a greater resistance to bending. Floor finishes are frequently made from timber (e.g. parquet flooring). This provides a long lasting surface with a natural appearance and a 'warm feel'.

Roof For the reasons already discussed in the previous paragraph, timber is widely used for the construction of the pitched roof. An important point to mention here is that because timber is a very variable material it needs to be **stress graded** when used for structural purposes. This means that each piece must be classified according to its strength. Stress grading is necessary because the strength of two pieces of timber even from the same species and tree can vary substantially. Variations in strength are caused by the presence of knots, the slope of the grain and growth rate of the material. Timber can be graded visually by examining each piece by eye. Alternatively, machine stress grading can be used. In the latter, the *stiffness* of each piece of timber is tested and related to the strength.

Window frames As previously discussed, there are a number of suitable materials for window frames and sashes. Timber is especially suitable because of ease of availability. This material is also very easy to shape into the required frame sections.

However, there are various problems associated with timber frame windows. Because the window is exposed to the external environment, the frames become wet when it rains. This means that moisture is absorbed by the window frames and window sills. As a result, fungal decay may occur. Decay is far more likely where sapwood instead of heartwood has been used in the manufacture of the window frame. It is also important to use timber which has a natural resistance to decay, i.e. has high durability when exposed to the weather. Hardwoods have greater durability but are more expensive. Timber species for use in window frames can be Douglas fir (softwood) or mahogany (hardwood), although the latter is more expensive. Species with poor durability such as beech must not be used for window frames.

> **Self-assessment task 2.41**
>
> As we have seen above, high levels of moisture in timber can cause rotting of the frame. What other problems are associated with varying moisture levels in timber?

Timber-framed houses Traditionally, houses in the UK have been built from bricks and blocks. However, timber-framed houses have various advantages. They are relatively lightweight, strong structures and have the great advantage of fast erection. In the UK, most timber-framed houses are made from factory assembled open-framed sheathed panels. Stability and rigidity are obtained after the panels have been bolted to the foundations and also joined to each other. Timber-framed houses often have brickwork as external cladding.

> **Self-assessment task 2.42**
>
> Sitka spruce (home grown) and Western hemlock (imported from North America) are two timbers used structurally in construction. Try to visit a timber yard and find other timber species which are in common use in building.

Timber products

These are materials whose main ingredient is timber. Timber products include **plywoods**, **particle boards** and **fibreboards**.

Plywoods

Plywood is produced by combining several layers of timber with adhesives. Each successive layer is at right angles to the previous one so that the strength properties are similar in the two respective directions (unlike solid timber). Figure 2.49 shows several types of plywoods. Can you remember why there is always an odd number of layers of veneer in plywood? (If not, read page 73 again.)

Advantages of plywood over solid timber are:

- stronger and stiffer
- resists splitting and so can be nailed or screwed near the edges

Plywoods can be used structurally for timber-framed house construction. Another structural use is for flooring.

Particle boards

Particle boards are also known as chipboards. (Standard grade chipboard should be very familiar as it is used frequently for bedroom furniture.) Particle boards are made from (waste) particles of wood of various sizes. The particles of wood are arranged randomly and are bonded together with thermosetting resins (see page 86). Advantages in using chipboards are:

- low cost
- chipboards have very little moisture movement
- unlike solid timber the strength is the same in all directions (why?)

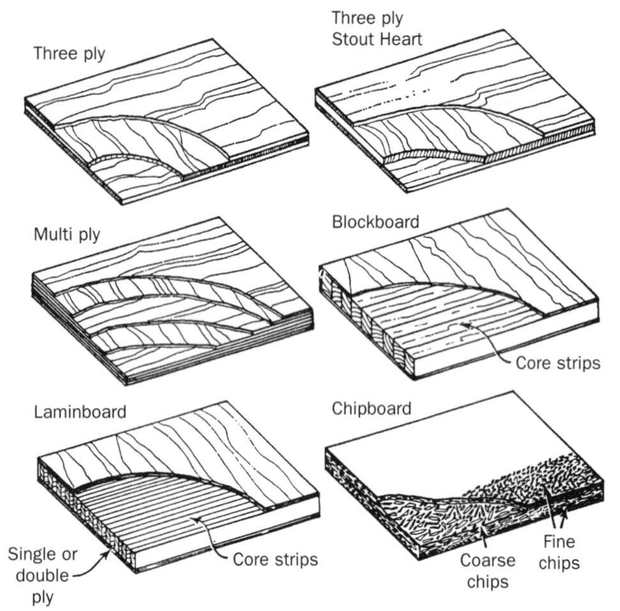

Figure 2.49 Various types of manufactured boards including several types of plywood and chipboard

A major disadvantage in using this material is that standard chipboard (type I) distorts and deteriorates quickly when exposed to damp conditions. Where dampness can be a problem, for example in bathrooms, moisture-resistant boards (type III) should be used.

Relevant uses in building are as follows:

- flooring (type II or type III chipboard should be used)
- roofing: can be used either for flat roofs or as roof boards (sarking) for pitched roofs. The roof tiles are placed above the sarking.

Fibre building boards

Fibre boards are composed from compressed wood fibres. The more the fibres are compressed, the greater the density becomes. The material is held together by the fibres themselves – binders are not usually added (compare with particle boards). Fibre boards include:

- **Softboards.** In this material the fibres are loosely bound. This gives the material a low density (e.g. 300 kg/m^3). Softboards are good thermal insulators (why?) and for this reason they are used for wall covering and on floors, underneath the floor coverings.
- **Medium boards.** As the name suggests, these boards have medium density (typically 700 kg/m^3). They are harder and have greater strength than softboard. Uses are similar to those of softboard.
- **Hardboards.** The particles are highly compressed, so this material has higher density than other fibre boards (around 900 kg/m^3). Among other uses, hardboard is employed as low cost lining, i.e. cover for partitions.

Self-assessment task 2.43

What are the main advantages of chipboard over solid timber? Where in the building structure is chipboard utilised?

Clay products

Clay products include bricks, blocks, cladding tiles and roof tiles. **Bricks** are frequently used (with mortar) in the construction of load-bearing walls.

Brick walls are

- durable
- quite dense
- strong in compression
- rigid (high stiffness or Young's modulus)
- water resistant

Bricks can be put into various categories:

- **Engineering bricks** are hard and dense. As we have seen previously, they have very high compressive strengths and low water absorption. These bricks are suitable for use underground and below the damp-proof course because the moisture absorption is low.
- **Common bricks** are suitable for general building work but are not always attractive enough to be used for facing.
- **Facing bricks** are selected and used for their attractive appearance on exterior walls. A variety of colours and textures are available.

The use of bricks for external walls to houses has the following additional advantages:

- people find that brick walls have an attractive appearance
- low maintenance costs
- because of the small size of bricks, almost any design can be built using this material

Certain types of **building blocks** are made from clay. Blocks are building units which are larger than bricks. Clay blocks are generally hollow which helps to make them reasonably light and also increases the thermal insulation (why?). The blocks can be used for the inner leaf of external walls and also for partition walls. Figure 2.50 shows several typical clay blocks. Why are the surfaces of the blocks keyed?

Many types of **roof tiles** are made from clay as they are very resistant to water penetration. The angle of the roof also ensures that water is shed from the surfaces quickly. The tiled roof surface must be strong enough to withstand loads caused by snow or by persons carrying out maintenance.

Tiles made from clay are frequently used on front facing external walls of houses. Figure 2.45 shows a domestic property where the tiling forms the 'club pattern'. These tiles can be used for decorative purposes as well as providing an adequate barrier to the weather.

Self-assessment task 2.44

Give two reasons why bricks with frogs or perforations are preferred to 'solid' bricks in construction.

Concrete

Concrete is used for the foundations of the house. It is straightforward to produce concrete of the required strength. Fresh concrete can be cast into the required size and shape as needed. The concrete used for foundations may be non-reinforced or reinforced, depending on the requirements and the grade needed. Non-reinforced concrete is plain – no

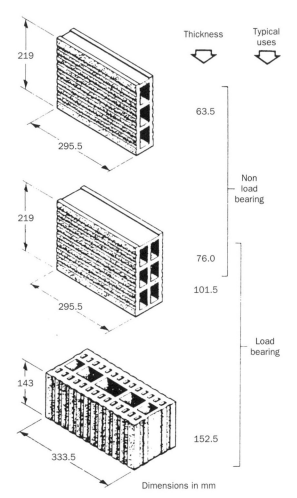

Figure 2.50 Several clay building blocks

strengthening materials are included. Reinforced concrete has steel bars or steel mesh embedded inside in order to increase the strength.

Table 2.14 Several grades of concrete and their uses

Grade	Characteristic strength at 28 days (N/mm²)	Lowest grade used
C7.5	7.5	Plain concrete
C15	15	Reinforced concrete using lightweight concrete
C20	20	Reinforced concrete using dense concrete

Table 2.14 shows several **grades** for concrete. As an example, grade C7.5 refers to concrete of characteristic compressive strength 7.5 N/mm² at 28 days. The word characteristic means that the stated strength (in this case 7.5 N/mm²) is exceeded by a specified percentage of samples, usually 95%. As an example, assume grade C15 has been specified. In this case, 95 in every 100 samples tested should have a compressive strength of at least 15 N/mm². Note from Table 2.14 that C7.5 is the lowest permitted grade for plain (or non-reinforced concrete). The minimum grade for reinforced (dense) concrete is C20. In other words, concrete of characteristic strength *less than* 20 N/mm² should *not* be used for reinforced concrete.

Self-assessment task 2.45

Grade C10 concrete was specified for the foundations of a certain house. Explain in your own words what 'C10' means. If 50 concrete cubes of this grade are tested, how many would you expect to have a strength *less* than 10 N/mm²?

Concrete is also used for 'solid' ground floors. The thickness of the concrete slab is typically 100 mm to 150 mm. Steel mesh reinforcement is usually required for larger areas.

External to the house, concrete paths and drives are commonplace. These are produced usually from non-reinforced concrete.

Concrete products

Concrete blocks are often used for the inner leaf of external walls. They are also employed for partition walls. The blocks may be solid or hollow. Concrete blocks can be low density or high density. Some important points follow:

- Many materials can be used in the manufacture of concrete blocks. Some of the materials are by-products from other processes. Materials used in the manufacture include natural aggregates and blast-furnace slag (this material is 'waste' produced from a process used to manufacture iron).
- Walls built with concrete blocks are considerably cheaper than brick walls. (Why?)
- High-density concrete blocks are strong enough for load-bearing walls.
- Low-density concrete blocks are important because they are good thermal insulators. These blocks are often aerated during manufacture.

Self-assessment task 2.46

Try to explain scientifically why low-density concrete blocks are better thermal insulators than high-density blocks.

Certain types of roof tiles are made from concrete. As with tiles made from other materials, rain penetration is minimised by overlapping the tiles. The shape of the tiles minimises capillary action.

Metals

Ferrous metals

Ferrous metals contain iron. Steels and cast irons are examples of ferrous metals.

In the building industry, (mild) steel is frequently used for structural purposes. This is because:

- steel is very strong in tension, compression and in bending
- steel has a very high Young's modulus which makes the structure rigid

Cast iron has various uses such as waste pipes, rainwater pipes, manhole covers, baths, boilers and radiators for central heating, etc. The use of cast iron has been reduced in recent times because many alternative materials are now available. For example, plastic pipes have largely replaced cast iron pipes.

Figure 2.51 The frame of this warehouse is constructed from structural steel. Notice the cladding panels in the foreground. These are made from plastic-coated steel, encasing expanded polyurethane for thermal insulation

---

Self assessment task 2.47

In terms of the physical and chemical properties, give three advantages in using plastic over cast iron for rainwater pipes.

Industrial and commercial buildings often have steel frames. Figure 2.51 shows a new warehouse under construction. The skeleton is being produced from steel. In domestic buildings, structural steel can be used for load supporting beams. Lintels are beams that carry the loads above doors and windows. (They are usually not visible and are hidden behind the brickwork.) Lintels are usually made from cold-rolled steel and are often known under the trade name Catnic. Steel lintels are usually galvanised (why?) and coated with a chemical-resistant paint. Lintels can also be made from concrete. The concrete used for such purposes *must* be reinforced with steel bars. As we have previously seen, bending forces cause part of the beam to be in tension. Because concrete is weak in tension, the steel adds great strength to the material in this mode.

Another type of load supporting beam is the rolled steel joist or RSJ. Suppose that part of a load-supporting wall is to be removed because an extension is to be built on to a house. In this case, an RSJ is substituted for the wall; this is the load-carrying beam. The universal steel beam is easy to recognise as it is 'I' shape in cross-section. The castellated steel beam is also 'I' shape – this is recognised by the large hexagonal holes which are symmetric about the middle of the beam. This provides a deeper beam for the same weight of material as the universal beam. The castellated beam can be used where deep floors are needed.

There are many other uses for steel in domestic buildings. Steel wall ties are used to connect the inner and outer leaf of external walls. This improves the stability of the structure. It is vital that these components do not corrode. For this reason wall ties are often made of galvanised steel or stainless steel. Mild steel straps are used to secure the timber roof trusses to the wall.

Non-ferrous metals

Non-ferrous metals do not contain iron. Aluminium, copper and lead are examples. (Zinc is another example which was previously mentioned. Because of its durability, it is commonly used for galvanising; the zinc forms a protective layer to stop steel rusting.)

Aluminium is used for gutters and flashings. (See below for the meaning of flashing.) Because aluminium has low density and has high corrosion resistance, it is ideal for these purposes. A layer of aluminium foil on the surface of insulation boards greatly reduces heat loss by radiation. Aluminium foil forms a very effective vapour barrier (see page 58) as this material has superb resistance to the passage of moisture and water vapour.

Aluminium alloys contain other metals to make the material stronger. Lightweight window frames made from aluminium alloys are very durable. (The window frames of the house in Fig. 2.45 are aluminium alloy, with hardwood surrounds.) However, they are more expensive than frames made from timber or uPVC.

Copper has very good corrosion resistance. It can be used for domestic roofing but is usually associated with roofs for large buildings. Because copper is easy to bend and shape (malleable) it is commonly used for service pipes, e.g. to carry water. Another property of copper, its high electrical conductivity, makes it invaluable for electrical cables.

Lead is a very soft metal – it is easy to bend into shape. It is very durable and this makes it suitable for weatherings and flashings. Weatherings are sloped external surfaces which are designed to shed water. Flashings are used for waterproofing the joints between two external surfaces. Figure 2.45 clearly shows lead flashing just below the first floor windows – this flashing is bridging the gap between the window and the tile wall cladding.

Self-assessment task 2.48

There is more aluminium in the earth's crust than any other metal. Why, then, is it such an expensive building material?

Bituminous materials

Bitumen is a black material which can occur naturally but the majority is produced from petroleum. Because of its low stiffness (small Young's modulus) it is rarely used on its own but is combined with other materials depending on the specific use. Some of the other important properties are as follows:

• It is brittle at low temperatures. (Bitumen softens as the

Figure 2.52 Various types of roofing felt

Figure 2.53 It can clearly be seen that even the combustion of some (plastic) videotapes produces a substantial fire and much smoke. Luckily, this was just a scientific test

temperature is increased but as bitumen is composed from a variety of molecules there is no definite melting point.)
- It is waterproof.
- It has good adhesion properties.

Bitumen is used to manufacture various kinds of asphalt. The type known as mastic asphalt is frequently used in building. Mastic asphalt contains bitumen and filler, which is usually powdered limestone. It needs to be heated to temperatures around 200 °C before application. This material is used for

- flat roofing
- damp-proof courses
- road surfaces

Self-assessment task 2.49

Figure 2.52 shows various types of roofing felts. What are these materials often manufactured from? What important property do these materials have?

Plastics (thermoplastics and thermosetting plastics)

There are two categories: thermoplastics and thermosetting plastics. Thermoplastics become softened by heating and can be moulded or remoulded as desired. Thermosetting plastics cannot be softened by heating; they must be moulded into the required shape during manufacture. The important desirable properties of plastics are:

- low density
- non-porous
- frost resistance
- unaffected by many chemicals
- good thermal resistance
- virtually any shape can be produced

Undesirable properties are:

- high thermal movement
- low Young's modulus
- plastics are combustible and produce poisonous gases when they burn (see Fig. 2.53)
- ultraviolet radiation from the sun degrades many plastics

As mentioned earlier, plastics are produced from fossil fuels, consuming large quantities of non-renewable resources. This is another disadvantage.

Thermoplastics

Several common thermoplastics and their uses for the building elements are now briefly discussed.

Polyethylene (PE) The uses of polyethylene include:

- water pipes
- cold water cisterns
- damp-proof courses (dpc) and damp-proof membranes (dpm)

For long lengths, polyethylene pipes must have a 'kink' to allow for thermal movement. Polyethylene is not suitable for hot water pipes for the same reason.

Polyethylene does not absorb water and it is not porous. For this reason it is suitable for damp-proof courses and damp-proof membranes. The lifetime of a polyethylene dpc is extremely long.

Polyvinyl chloride (PVC) In its 'normal' form this plastic is rigid – it is called *unplasticised* polyvinyl chloride or uPVC. Unplasticised PVC has good resistance to chemicals such as acids or alkalis and is hardly affected by the weather. This makes it suitable for:

- pipes for water mains
- soil and waste pipes
- drains and sewers

Further, its electrical insulation properties make uPVC very suitable for electrical accessories. The big disadvantage of plastics generally is high thermal movement and low rigidity. However, uPVC windows have been quite successfully used domestically for many years. Frames made from uPVC have the great advantage of having low thermal conductivity (compare with steel or aluminium windows). This property reduces problems of condensation onto the window frame.

Plasticised polyvinyl chloride This form of PVC is 'flexible' because plasticisers have been added. Uses include:

- sarking for roofs
- floor coverings
- insulation for electrical cables

Expanded polystyrene (EPS) Polystyrene is a transparent plastic with an appearance which resembles glass or Perspex. It has about the same density as water. On the other hand, *expanded* polystyrene is white and has a very low density. Most of the space taken by the material is in fact air. This makes the material a very good thermal insulator. Unfortunately, EPS burns easily and great care must be taken in its use. It is used for the thermal insulation of:

- cavity walls
- ground floors
- roofs

Self-assessment task 2.50

The many other thermosetting plastics used in building include polyvinyl acetate (PVAC), polypropylene (PP) and polytetrafluoroethylene (PTFE). Find out the uses for these materials together with their relevant properties.

Thermosetting plastics

Thermosetting plastics are normally produced by mixing a *resin* with a *hardener*. A chemical reaction occurs and the resulting material then sets. Thermosetting plastics cannot be softened by heating. These materials are hard and rigid compared with thermoplastics. Some examples of thermosetting plastics and their uses follow:

Phenol formaldehyde (PF or Bakelite) This was the first thermosetting plastic to be produced. Uses include electrical applications (it is a very poor conductor of electricity). It is also used in adhesives and paints.

Polyurethanes (PU) Polyurethanes are used in the manufacture of paints and varnishes. Foamed polyurethane has a very low density (around $24 \, \text{kg/m}^3$) and is an excellent thermal insulator. A major problem is that foamed polyurethane is highly combustible, evolving poisonous gases as it burns. It should not be used where direct contact with fire is possible. It is often injected into the cavities of external walls to increase the thermal insulation.

Glass-reinforced polyester (GRP) Polyester resins on their own are brittle and have a low value of Young's modulus. This material can be **reinforced** using glass fibres which are much stiffer. This produces glass-reinforced polyester which is tougher (more resistant to impact) and much stiffer than the polyester resin on its own. The uses of GRP include pipes for water supply or sewers, and precast moulded products including baths and shower cubicles.

Self-assessment task 2.57

1. Apart from GRP, can you think of another building material which is commonly reinforced? What is the reinforcing material in this other product and why is it necessary?
2. The foundations of dwelling houses are usually constructed from which material?
   (a) zinc
   (b) mild steel
   (c) aerated concrete
   (d) plain concrete
   (e) wood
3. Which **one** of the following is the most suitable material for the rafters of a pitched roof?
   (a) medium-density board
   (b) clay blocks
   (c) preservative-treated softwood
   (d) expanded polystyrene
   (e) lead
4. Which **one** is the most suitable material for use as loft insulation?
   (a) polythene sheeting
   (b) hardboard
   (c) uPVC
   (d) plasterboard
   (e) glass fibre/mineral wool quilt
5. A beam is required to span a 3 m wide opening. Which is the most suitable material for the beam?
   (a) steel
   (b) Perspex
   (c) clay
   (d) in-situ plain concrete
6. The damp-proof membrane in the ground floor of a house is usually made from
   (a) chipboard
   (b) plasterboard
   (c) polythene sheeting
   (d) mild steel sheeting
   (e) hardboard
7. Which material can warp and become distorted if there is considerable change in the moisture content?
   (a) timber
   (b) glass
   (c) bitumen
   (d) copper
8. Which material is strong in compression but weak in tension?
   (a) structural steel
   (b) plain concrete blocks
   (c) reinforced concrete
   (d) timber

Conclusion

The science of materials is a fascinating subject. As we have seen, the *properties* of our building materials are most important in deciding their uses. However, the appearance, availability, cost and environmental considerations must also be taken into account when choosing materials.

Construction Technology and Design

Des Millward

This unit introduces the principles and practices of the construction process applied to low-rise buildings. The unit encourages consideration of the relationships between the function of a building and its structural form, together with how the internal division of space is achieved. Throughout there is an emphasis on the performance of the building and its many parts. Attention is then drawn to the provision and installation of building services which contribute towards the maintenance, health, comfort and convenience of the building user. Finally the unit considers some of the basic operations associated with the external works for a residential development. Throughout the unit there is emphasis on the application of the various health and safety legislation, regulations and British Standards. This unit uses these themes to introduce the reader to:

- Relationships between design and function of low-rise buildings
- Construction of low-rise buildings
- Primary services in low-rise domestic buildings
- External works for a residential development

3.1 Relationship between design and function of low-rise buildings

The environment in which we live, work and relax is made up from a mixture of elements which are naturally occurring and those which have been created by man. These are usually known as the:

- **natural environment**
- **built environment**

The **natural environment** includes grasses and wild flowers together with trees, shrubs and bushes which are located upon the earth's surface. The earth's surface has also been modified by natural forces such as the sun, wind, rain and geological movement. This has created features such as rivers and lakes, valleys and hills which have continued to influence the location and development of the built environment.

The **built environment** since early times shows how the human race sought to modify the climatic conditions in order that particular needs were satisfied. These needs have varied over time as the sophistication of society's demands have increased. Often the development of the built environment has sought to retain and integrate the features of the natural environment.

Today's buildings are easily identified by their use. Planning authorities relate the use of a building to the area in which it is situated. These areas are known as zones which normally for example avoid having residential areas within industrial areas. This is a feature of our environmental planning. Use classes are identified by planning authorities and are:

- commercial uses
- business and industrial uses
- residential use
- non-residential institutions and leisure uses

These use classes contain buildings which we know as:

- shops, banks, estate agents, restaurants
- offices, factories, warehouses
- houses, hotels and residential care homes
- sporting complexes, schools, libraries and churches

These buildings all create the need for:

- a structural form which will enclose a space
- an internal environment providing conditions appropriate to activities
- maintaining the comfort and safety of the occupants

Function of low-rise buildings and their structural form

Buildings pass through a design process before they can be constructed. This process needs to establish certain basic facts relating to the proposed building. Two kinds of information will need to be incorporated into the design:

- client and user requirements
- legislative requirements applied to the construction of the building

These influences will control the design of the building when the basic arrangements for spaces, their use and volume will be determined. This leads to the development of the basic form of the building which in turn leads towards the selection of a particular structural form.

While different structural forms may be used to accommodate user requirements the broad classification of the building use will be common, for example:

- **residential accommodation** – houses or blocks of flats
- **storage of goods** – furniture, food, timber, electrical equipment
- **commercial** – offices, shops

Legislative requirements can relate to the design, construction and operation of a building. Examples where requirements are stated are:

- Building Regulations and Approved Documents
- British Standards
- Water Byelaws
- Institute of Electrical Engineers Wiring Regulations
- Gas Safety (Installation and Use) Regulations
- Town and Country Planning Act

Structural forms

Structural forms are used to create a building enclosure and carry the various loads. There are two general classifications for structural form:

- masswall
- framed

Masswall buildings are traditional in construction and use brick, block, stone or concrete walls to carry the structural load. In addition the walls can also perform other functions such as:

- excluding weather
- providing security
- thermal, sound and fire resistance

Framed structures as the name suggests create a structural framework within which operations relating to the building use can be accommodated. Initially framed structures are not able to perform the other functions listed for masswall. This is because the frame can only carry structural load. Some form of cladding is required to enclose the frame which can then satisfy those functions. Table 3.1 classifies the structural forms.

Table 3.1 Typical structural forms

Masswall	Framed
Monolithic	Shed
Cellular	Skeletal
Crosswall	Portal
Timber frame	
These are illustrated in Fig. 3.1	

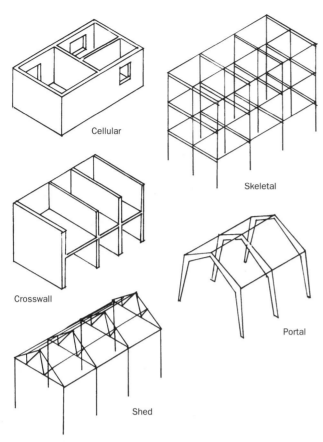

Figure 3.1 Structural forms

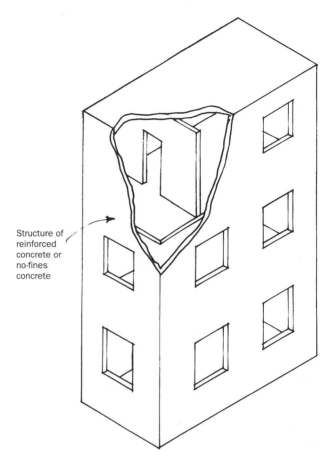

Figure 3.2 Monolithic structure

Structural forms for low-rise buildings

Buildings are often loosely classified by the number of storeys as:

- low-rise – up to three storeys
- medium-rise – up to eight storeys
- high-rise – beyond eight storeys

In addition to being categorised by height, buildings may also be constructed using particular structural forms. These forms can be related to the use to which the building is to be put initially. It is worth remembering that many buildings throughout their life undergo internal space reorganisation to meet the changing needs of the occupier, thus allowing a building to be recycled, possibly falling into a different use class.

Masswall structural form

These structures use solid walls to define the enclosing space within the building. The walls receive, carry and transmit the structural load of the building safely to the foundations and the ground beneath.

Monolithic structures are those which use concrete as the medium for carrying structural loads (Fig. 3.2). They are cast in-situ and may be either reinforced dense concrete, or no fines concrete. The walls, and in some cases the floors being cast at the same time, serve to increase the structural stability of the building. Common applications are blocks of flats and maisonettes.

Cellular structures are similar in form to monolithic structures, the exception being that they are constructed using bonded brickwork. Internal walls are used to divide the spaces and thus form self-contained areas. Some of the internal walls may be load bearing and therefore the external walls may not carry the total structural loads of the building. Common forms are houses (Fig. 3.3), bungalows, terraced housing, flats, care homes.

Crosswall structures use a series of parallel walls which carry structural loading from the floors and roof down through the foundations to the ground. Internal space is often divided up by using non-load-bearing block, or timber stud partition walls. Common applications are for terraced housing (Fig. 3.4), or combining residential accommodation with retail shops (Fig. 3.5).

Framed structures

Framed structures (Fig. 3.6) use timber, steel or reinforced concrete in a variety of shapes to carry compressive and tensile structural loads safely through the foundations to the ground. The enclosing and dividing structure between the building frame need not be load bearing, but in some cases it may be necessary to provide additional support to maintain stability of the frame.

Timber frame

This is a particular form of framed construction which rarely exceeds three storeys in height. Timber panels are manufactured off site and assembled on site on a pre-prepared foundation and sub-base (Fig. 3.7). The external enclosure and internal sub-division of space is created with timber frame. Various forms of cladding to the frame are used, some of which increase rigidity of the frame. Typical applications are for housing, small offices and residential care buildings.

Figure 3.3 Cellular structure

Figure 3.5 Crosswall residential and retail property

Figure 3.4 Crosswall terraced housing

Figure 3.6 Steel-framed structure

Skeletal structures

A general term used to describe a framed rectangular structure using columns and beams as the main load-bearing elements. Steel and reinforced concrete are common, though laminated timber has also been used (Fig. 3.8). Applications are for offices, shops, flats, hotels, and residential care buildings.

Portal frames

These structures have no internal columns which allows large clear internal floor areas to be created, also high storey heights are possible. The steel or concrete frame consists of spanning and supporting members and is available in many forms (Fig. 3.9). The frames may be used alongside each other to form multi-bay construction. Applications include warehouses, factory units, manufacturing plants.

Shed structures

These are similar to portal frames in that they form large clear internal floor areas, however they differ in that the span between the supporting members is filled by either braced girders or roof trusses (Fig. 3.10). Applications are similar to portal frames.

Self-assessment tasks

1. Consider a house, warehouse and an office block in your locality. Describe their appearance and structural form.
2. Describe the materials used in their construction and assess their load-carrying abilities.
3. Explain the essential difference between a cellular and a crosswall structural form.
4. Draw simple sketches to show the difference between a portal frame and a shed structure.
5. Prepare a collage to illustrate 10 different building types.

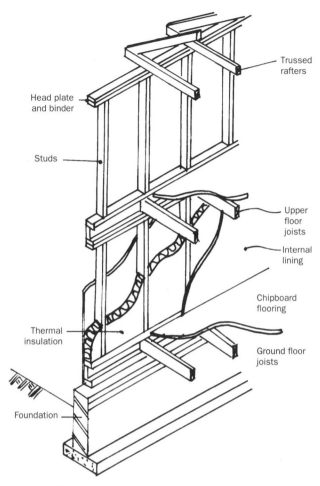

Figure 3.7 Timber frame construction

Figure 3.8 Skeletal structure – office block

Figure 3.9 Portal frame – warehouse

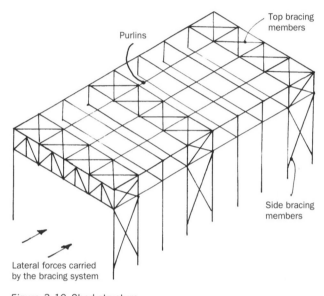

Figure 3.10 Shed structure

Internal division of space related to building function

Residential accommodation

The use of space in houses, hotels and residential homes often creates areas that are used for different purposes, for example:

- kitchen
- bathroom
- lounge
- bedroom
- circulation space

In general the spaces are often no larger than 5.0 metres in length or width and 2.3 metres storey height. These spaces are created by the use of internal walls and floors, which allow monolithic, cellular, timber frame or crosswall construction to be used. For these applications it is important to recognise the influence of:

- circulation or access routes
- the relationship of building services, particularly sanitation and drainage, to the internal layout of the building
- fire escape routes

- disabled access
- position of noisy and quiet areas
- orientation of the building
- ventilation requirements
- natural lighting

A range of dwelling types is available that will vary both in size and physical characteristics. These include:

- Flats (Fig. 3.11)
- Maisonettes (Fig. 3.12)
- Bungalows
- Houses (Figs 3.3, 3.13, 3.14)
- Sheltered accommodation
- Residential care houses

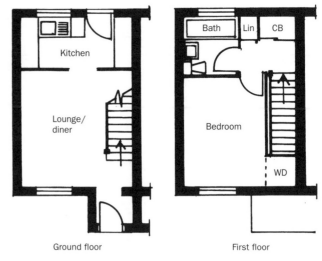

Ground floor   First floor

Figure 3.13 Terraced one-bedroom starter home

Figure 3.11 Low-rise block of flats

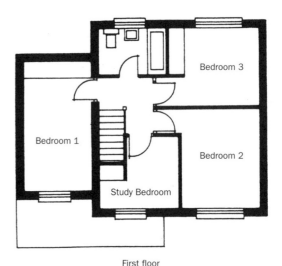

First floor

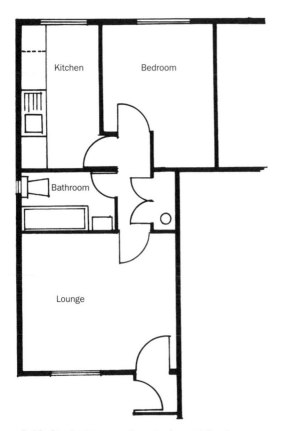

Figure 3.12 One-bedroom maisonette (ground floor)

Ground floor

Figure 3.14 Four bedroom detached house

Closer examination of the floor plans for a **starter home, maisonette** and **detached house** will reveal a number of common features that a designer has considered. These have been interpreted in different ways to satisfy the anticipated needs of potential occupiers.

All of the dwellings are surrounded by a structure that provides a climatic barrier to enable an internal environment to be maintained to meet certain criteria, the principal one being that of temperature. The **Building Regulations** influence buildings in that they set minimum standards for those who construct buildings. The areas the regulations cover are:

- Structure
- Fire safety
- Site preparation and resistance to moisture
- Toxic substances
- Resistance to the passage of sound
- Ventilation
- Hygiene
- Drainage and waste disposal
- Heat producing appliances
- Stairs, ramps and guards
- Conservation of fuel and power
- Access and facilities for disabled
- Glazing – materials and workmanship
- Materials and workmanship

The **external walls** of buildings will have to be able to:

- remain structurally stable
- resist the passage of moisture, heat and sound
- prevent the passage of fire

The resistance to the passage of sound and fire is more important in walls of:

- terraced housing
- flats
- maisonettes
- semi-detached houses

Walls between the dwellings known as **compartment walls** will meet specific requirements for sound and fire. Internal walls or **partitions** normally serve the function of dividing spaces for particular uses. These spaces are clearly identified on the building plans.

Designers recognise the importance of providing natural light to spaces. This is achieved by placing windows in strategic positions to provide an acceptable level of daylight. In general the larger the window area the higher the lighting level and risk of glare from sunlight. It must be recognised however that windows are one major area of heat loss in a domestic property and so their size and area may be restricted. **Double glazed** windows will assist in reducing the amount of heat lost.

Habitable spaces are ventilated either by **natural or mechanical ventilation**. Ventilation being a means by which odours and moisture vapour can be controlled. Opening windows are used to meet the minimum requirements. Principal rooms in domestic properties requiring ventilation are:

- kitchens
- bathrooms
- toilets
- en-suites
- shower rooms

Use of space in a property will vary according to the time of day and the number and type of occupants. **Circulation space** allows building users to enter a property and access the various rooms within it. The circulation space will vary according to the type of property.

One-bedroom starter homes and flats (Fig. 3.13) are compact in design often combining access, living and circulation space. This often involves passing through the lounge area to be able to reach the kitchen and upper floors. Larger detached properties incorporate entrance lobbies, hallways, open staircases and extensive landings (Fig. 3.14).

Where dwellings are designed specifically for **disabled people** special consideration should be given to:

- location and position of sanitary accommodation
- size of accommodation to allow wheelchair access and circulation
- height of fittings and support rails (door furniture, electrical, sanitary)
- external access
- thermostatically controlled mixer taps
- operations relating to water supply and discharge from fittings
- height of fittings – ramps, hand and guard rails

Likewise the use of a space in smaller properties may combine uses, for example:

- kitchen/diner
- lounge/diner
- living room/kitchen

Where possible the building should be designed such that water and drainage supplies are able to be installed in an economic yet functional manner. This involves the designer thinking through the location of:

- kitchens
- bathrooms
- utility areas
- en-suites
- cloakrooms

In the case of dwellings contained on the same floor level a designer will seek to position them alongside each other as illustrated (Fig. 3.12) or back to back. In the case of houses the practice is to attempt to place kitchens below bathrooms (Figs 3.13, 3.14) to concentrate hot and cold water supplies and to enable the installation of a soil stack. The stack can then serve a variety of appliances in removing waste products.

In blocks of flats, kitchen and bathroom areas are placed side by side within a dwelling and back to back with another dwelling within a horizontal zone. In the vertical zone the kitchens and bathrooms will be placed immediately above each other on every floor to enable the use of **service cores**. These will accommodate water and drainage services.

The reader should bear in mind that there is not one optimum layout for a particular dwelling type, rather that the final layout seeks to be functional and cost effective.

Storage of goods

Warehouse owners often require a functional, low cost and maintenance-free building in order to satisfy their key requirements of large clear floor areas and high ceilings. **Portal frame** and **shed** buildings (Fig. 3.1) are ideal for this purpose.

Economy of construction is possible by using steel or concrete to create the structural frame. These materials are ideal as they possess good strength to cross-sectional area ratio and are able to cross unsupported spans to leave floor areas clear. They are also able to be manufactured in different forms. Enclosure of the structure is normally by using any of the proprietary wall and roof claddings available.

Internal layout will be dictated by the entry and exit points and the position of the stored goods. If mechanical handling is to be used access routes will need to be established and maintained.

Warehouses are often associated with factories as the goods produced will need to be stored before distribution. The key features associated with a warehouse are:

- large uninterrupted floor areas
- higher than normal storey heights
- good access through wide and high doorways
- acceptable lighting levels
- security
- provision for overhead craneage
- provision for containing fire

Examples of buildings suitable for storage space are shown in Figs 3.9 and 3.10. Key design features are:

- clear floor area
- maximum operational height
- access and egress

Figure 3.15 shows a portal frame warehouse. Clear floor area is dictated by the spacing of the structural frame, in this case the portal frames. Maximum clear height for transportation of goods is limited by the height of the eaves or the intermediate valleys in a multi-bay portal frame.

The position of the structural frame will dictate internal circulation space relative to the external doorway positions. This will be further compounded by the access required for fork lift trucks to stack and transport goods within and to the outside of the building. Figure 3.9 shows a warehouse with wide and high roller shutter doors to allow vehicular access to the building.

The provision of natural light is not a key requirement. The lighting levels required are those consistent with providing a safe working environment to satisfy the requirements of the

Health and Safety at Work etc. Act. While roof lights can be provided within the roof structure the useful light provided by them is likely to be inhibited by the high stacking of stored goods. Artificial lighting is therefore likely to be the primary light source.

Figure 3.10 shows a shed structure that will have a flat roof profile and constant internal clear height. This can provide additional usable height that a portal frame could not, and would lead to regular stacking of goods, and therefore an increase in storage capacity. Larger spans are also possible between columns which increases the flexibility of the storage space and is not as limiting as the portal on circulation space.

Some form of office accommodation is often required within a warehouse. The volume provided for administrative purposes will reflect the size of the commercial operation. Figure 3.15 shows such a situation. As the office accommodation is a separate activity to the storage of goods the office walls facing the storage area will have to be constructed as a fire **compartment** wall. In the event of a fire this will contain the fire within its area of origin, and possibly prevent it from breaking through into the unaffected area.

Commercial buildings

Businesses vary in size and structure and so does the internal organisation and layout of their office accommodation. Building structures used for offices must therefore be flexible in meeting the needs of the user. Small and large working spaces will be required which often means that internal partitions are rarely permanent or load bearing.

Skeletal frameworks are the most popular though **cellular** structures can be used (Fig. 3.1). Framed buildings are also able to accommodate air-conditioning and ventilation systems, telecommunication and electronic information systems. Internal division will mainly revolve around corridors, lifts, foyers and staircases.

The internal division of space in office buildings will vary from company to company. It is not unusual to find one company requiring large open plan offices with relatively clear floor areas, another requiring well defined work areas provided by permanent methods of internal space division, while a third would require a mixture of the two.

The designer is able by careful choice to meet these requirements by selecting appropriate structural grids for the building frame. In addition application of the long life, loose fit principle can enable future flexibility when re-designing the internal space.

Key features associated with offices are:

- comfort and safety of the occupants
- provision of daylight
- flexibility in the use of floor space
- ease of access about a floor of the building and between floors
- emergency escape routes in case of fire
- ability to accommodate, distribute and modify services

As previously mentioned, offices may be well-defined cellular work spaces or open plan. Figure 3.16 shows a detail of an office that combines both. Office buildings are most economical with a rectangular floor plan. The structural grid of the frame will therefore reflect this. The internal dividing walls will not normally be load bearing which leads to them being quite lightweight in construction, except in certain locations.

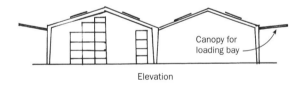

Elevation

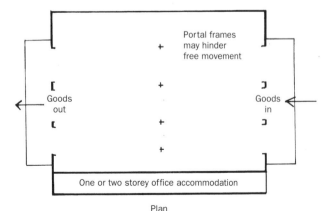

Plan

Figure 3.15 Traditional portal frame used for storage

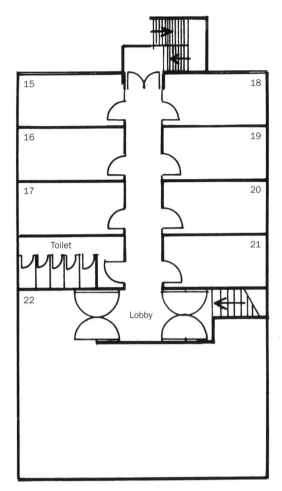

Figure 3.16 First floor plan of an office building

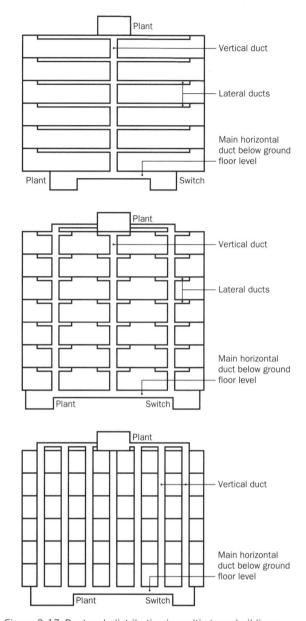

Figure 3.17 Ductwork distribution in multi-storey buildings

Natural lighting is an accepted feature of office buildings. Provided the internal depth of the room is not too deep then natural light can be used to good effect. The orientation of the building and window positions will influence the amount of light and solar gain admitted. This will also raise issues related to glare and sensible heat gain.

Vertical and horizontal circulation are important requirements in offices. Vertical movement is effected by the use of staircases and lifts. Both of these will require a fire resisting structure around them known as a **protected shaft**. This forms part of the **means of escape** in case of a fire. It should be noted that a lift should not be used in the event of fire.

Horizontal movement is linked to staircase or lift positions and normally will involve the use of a smoke lobby and door which link to a corridor, the corridor then providing access to the various working spaces and a **protected route** in the event of a fire.

It is unusual for the positions of staircases and lifts to vary from floor to floor and therefore their position dictates the vertical and horizontal access routes in the building. Further limitations on the flexible use of floor space will occur through the initial positioning of toilet facilities. As for blocks of flats, hot and cold water supply and drainage can be accommodated in service ducts which run vertically up the building. This will require toilet facilities to be located above each other throughout the building.

Other services now commonly associated with offices are those of mechanical ventilation and air-conditioning systems. Both these services require a plant room and provision within the building to accommodate ductwork

(Fig. 3.17). Again a service duct will accommodate vertical distribution, and suspended ceilings the horizontal ductwork distribution.

Hotels, schools and colleges may use similar floor layouts and access routes to office blocks. Building services require vertical and horizontal distribution zones and therefore dictate toilet and washroom facilities in a similar manner.

Figure 3.18 shows a floor plan for a butcher's shop. Public and private circulation spaces are required. The general public require access to the shop and also to be able to view the chilled display and approach the serving counter. This space should be large enough to accommodate a wheelchair user with ease.

Employees require access to store rooms and chill cabinets unhindered by the general public. Areas behind the serving counter and to the storeroom are therefore private or restricted access. The student should note the separate compartment for the WC and hand-washing facilities in order to reduce the risk of contaminating the food.

Shops are generally of single storey construction and designed principally to suit the convenience of the user. In

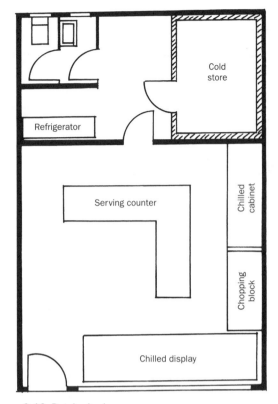

Figure 3.18 Butcher's shop

city centre sites where land is at a premium it is accepted practice to build offices on top of retail premises. In neighbourhood shopping areas it is usual to find flats or maisonettes built above shops (Fig. 3.5).

Construction costs

The cost of a building will be arrived at after various influencing factors have been taken into account. These will include:

- location
- cost of the land
- type of building
- shape of building
- amount and type of building services installed
- number of storeys
- demand for buildings

Comparative costs for a range of dwelling types are shown in Table 3.2. It can be seen from this the influence of space and height on building costs.

Table 3.2 Comparative costs of different dwelling types

Dwelling form	Cost index
Terraced bungalows	96
Two-storey houses	100
Detached bungalows	105
Two-storey flats	108
Three-storey houses	111
Three-storey flats	116
Four-storey flats	131
Five-storey flats	138
Other high-rise forms	140 upwards

Derived from Colquhon and Fraser (1991) *Housing Design in Practice*, Longman

Table 3.3 Typical unit rates per square metre of floor area (1996)

Building type	Cost/m^2
Factory workshop	500–800
Office block (low rise, non air-conditioned)	900–1180
Office block (medium rise, air-conditioned)	1100–1400
Shops (small)	420–535
Large department stores (including fitting out)	1050–1470
Four-star city centre hotel	1230–1470
Semi-detached house (estate location)	400–535
Detached house (bespoke)	600–820

For other types of building it is more difficult to illustrate typical costs because of the number of influencing variables. It is possible to show approximate unit rates per square metre of floor area to indicate a range of prices, but even within these there will be wide variations in practice. These can be used to forecast approximate building costs.

Self-assessment tasks

1. Identify and describe the information that influences a building design.
2. State five different structural forms for buildings and for each state two uses.
3. Describe the factors that will influence the internal floor plan of a:
   - house
   - warehouse
   - office
4. Prepare a 10 minute presentation on the factors which influenced the internal layout of a building known to you.

Materials used for building elements related to the building function

Buildings are made up from a series of parts which are called **elements** (Fig. 3.19). Elements are able to resist and carry different types of loads which assist the building to fulfil its purpose. The **primary elements** used in most forms of buildings are:

- foundations
- walls
- floors
- roofs
- frames

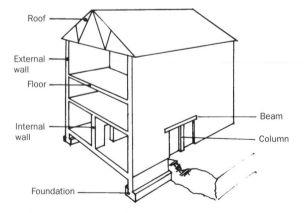

Figure 3.19 Elements of a building

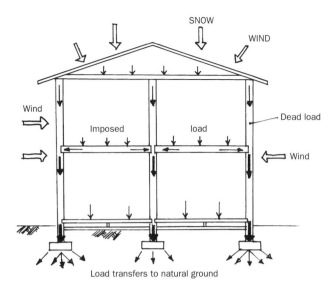

Figure 3.20 Loads on a building

The **loads** (Fig. 3.20) which one or more of these elements have to resist are:

- **wind**
- **snow**
- **dead**
- **imposed**

Major dead loads will include the self weight of the building structure and be made up of the load from the materials forming the primary elements. Imposed loads are a mixture of dead and live loads. Live loads can include people, and dead loads will include furniture, movable partitions and other equipment.

Designers know that certain things are expected of a building. These include:

- foundations should not sink into the ground
- the building should not fall down
- excessive movement should not occur
- floors should not deflect an undue amount
- roofs should not be blown off by the wind
- frames should not deform

Limits and conditions can be applied to each of these to suit particular circumstances, according to the type of building and its structural form. In other words the design will meet certain criteria for stated conditions. These criteria will in turn determine the structural **performance** of a building in use. Other criteria will be applied to the internal requirements.

In order to satisfy these performance criteria designers will need to relate the use of the building, the type of structure, and the materials used to each other. This in turn will affect the stability of a building. The appearance or aesthetics of the building may also need consideration. This is more so in the case of domestic properties and offices, though new cladding materials combined with different types of structural frames for industrial buildings can produce interesting profiles and colourful exteriors.

Factors which will affect the materials used to form the elements of a building are:

- loads to be carried
- height and length of walls
- clear span of roofs or floors

- degree of fire resistance required
- weatherproofing requirements
- aesthetics
- type of roof structure

The achievement of stability in low-rise buildings

Stability of buildings is often only associated with foundations, but the reader would do well to remember that stability applies to the whole of a building. The structural elements all integrate in assisting with a building's stability. Thus the structural elements play a large part in meeting the requirements of the Building Regulations by carrying, resisting and transmitting building loads safely to the ground.

Table 3.4 Characteristics and use of materials

Material	Element	Material characteristics
Brick	Walls	Variety of decorative surfaces
	Arches	Good load bearing qualities
	Chimneys	Standard modular sizes
		Good fire resistance
		Poor thermal qualities
		If high density resistant to moisture
Concrete	Foundations	Strong in compression
	Walls	Weak in tension unless reinforced
	Floors	If high density resistant to moisture
	Roof slabs	Good fire resistance
	Frames	Resists airborne sound
	Stairs	Poor thermal insulator unless air entrained
	Columns	Compatible with most materials
	Beams	Durable material
	Lintels	Attacked by sulphates
Steel	Frames	High strength to cross-sectional area
	Beams	Easily fabricated into different shapes
	Columns	Elastic material
	Lintels	Fire resistance good to 250 °C
	Reinforcement	Compatible with other materials
	Doors	Durable material
	Windows	
	Roof trusses	
Timber	Roofs	Easily fabricated into a range of components
	Trusses	Size varies with moisture content
	Floors	Durable if protected
	Partitions	Can be stress graded to determine load capacity
	Doors	Charring rate slow in fire which protects strength
	Windows	
	Stairs	

Foundations

Various types of foundation can be used. The choice of foundation depends upon the:

- soil conditions
- physical and legal restrictions on the site
- nature of the structure being supported

Foundations are constructed by using either mass or reinforced concrete. Mass concrete is used where the ground is able to support the foundation and prevent tensile stresses from causing movement or failure through shear. Two broad classifications for foundations are:

- shallow
- deep

Shallow foundations transfer their load within 3.00 metres of the ground level and deep foundations are beyond 3.00 metres. For economic reasons shallow foundations should be used where possible unless the situation demands otherwise. Shallow foundation types are:

- strip foundations
- pad foundations
- raft foundations
- short bored pile

Table 3.5 indicates the suitability of foundation types to soil and site conditions.

Table 3.5 Choice of foundation type in accordance with site and soil conditions

Soil and site conditions	Possible type of foundation
Rock or solid chalk, sand and gravels	Shallow strips or pads
Firm stiff clay with little vegetation liable to cause shrinkage or swelling	Strips 1 m below ground level or piles with ground beam
Firm stiff clay with trees close to the building	Trench fill or piles and ground beam
Firm stiff clays where trees have recently been felled and the ground is still absorbing moisture	Reinforced piles or thin reinforced rafts in conjunction with a flexible building structure
Soft clays or soft silty clays	Wide strips; up to 1 m wide or rafts
Peat or sites consisting partly of imported soil	Piles driven down to a firm strata of sub-soil
Where subsidence might be expected (e.g. mining districts)	Thin reinforced rafts

Source: Ashcroft *Construction for Interior Designers* Longman

For the sub-structure to perform satisfactorily in use it should be able to:

- safely transmit the dead and imposed loads from a building to the ground such that movement is limited and the ground is not over-stressed
- avoid damage from swelling, shrinkage or freezing of the sub-soil
- resist attack by sulphates or other harmful matter in the soil

Stability of the sub-structure

The stability of the foundation and the associated structure up to the damp-proof course is subject to:

- the ground conditions
- the construction of the sub-structure

The way in which the sub-soil behaves when it is under load can influence the behaviour of a foundation. Soils are classed under two broad headings:

- cohesive
- non-cohesive

Cohesive soils have a wide range of bearing capacities and are linked with long-term settlement. Clay is the major cohesive soil type which is subject to shrinkage and swelling through moisture changes. Typical non-cohesive soils are sand and gravel which have a high bearing capacity and lower settlement characteristics than cohesive soils. Typical ground-bearing capacities are shown in Table 3.6.

Table 3.6 Allowable ground-bearing capacities

Classification	Bearing capacity (kN/m^2)
Rocks	
Strong sandstone	4000
Schists	3000
Strong shale	2000
Granular soils	
Dense sand and gravel	>600
Medium dense gravel	200–600
Loose sand and gravel	<200
Compact sand	>300
Loose sand	<100
Cohesive soils	
Stiff boulder clay	300–600
Stiff clay	150–300
Firm clay	75–150
Soft clay and silt	<75

Adapted from: Table 1, BS 8004: 1986, reproduced by courtesy of BSI

In order to limit movement due to climatic changes **strip foundations** are normally a minimum of 900 mm below ground level (Fig. 3.21). **Deep strip foundations** (Fig. 3.22) and **short bored pile** (Fig. 3.23) extend deeper and beyond the clay or compression layers. **Reinforced strip foundations** (Fig. 3.24) are also used as an alternative to the deep strip method. **Raft foundations** are placed on the ground surface but their construction is such that they are able to accommodate shrinkage and swelling, subsidence and differential settlement (Fig. 3.25).

The depth to which foundations exert a vertical pressure can be determined by analysis of the **bulb of pressure**. This shows that the larger the loaded area the deeper will be the effect on the soil. Research has shown that consolidation of the soil beyond 20 per cent of the original load is negligible and therefore can be ignored. This is considered to be the effective bulb of pressure and can be quoted for various foundations (Table 3.7).

Table 3.7 Effective bulb of pressure depth

Foundation type	No. of times foundation width
Strip	3
Pad	1.5
Raft	1.4

Load transmission and strength

Load transmission and strength of foundations relies on:

- quality of the concrete
- thickness of the foundation

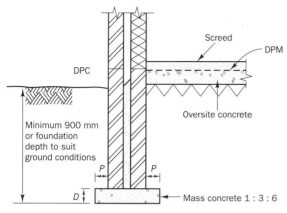

Depth *D* must be 150 mm minimum and must equal *P*.
P must be equal projection either side of the wall

Figure 3.21 Strip foundation

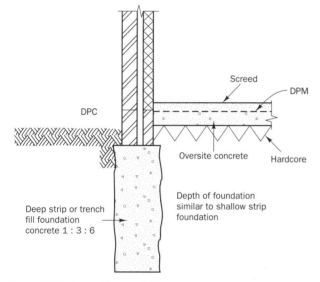

Figure 3.22 Deep strip foundation

- load-bearing ability of the ground
- type of structure
- type of foundation

The construction of the sub-structure has already been discussed earlier in the chapter. It is important to remember that with most types of foundations stability is increased when loads are placed centrally to the foundation to provide axial loading rather than eccentric loading.

Compatibility and durability

The successful performance of the foundation will depend upon the compatibility of the:

- foundation to the type of structure
- foundation to the ground load-bearing characteristics
- soil conditions to the foundation

The choice of foundation to match the structural load is an important one, as the transfer of the load must be effectively transmitted to the ground. Where the foundation is not matched, failure of the foundation through settlement or breaking up could be the result.

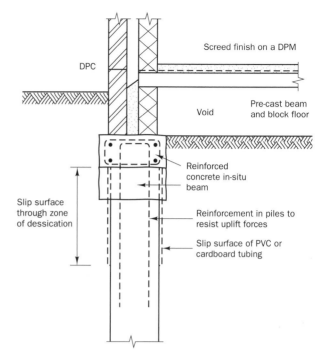

Figure 3.23 Pile and beam foundation to resist soil heave

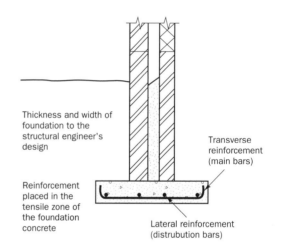

Figure 3.24 Reinforced wide strip foundation

Where the ground load-bearing characteristics are known, for example a weak 2.0 m layer of soil near the surface, then deeper foundations will be required such as short bored piles to carry the load to a suitable load-bearing strata. Again settlement of the foundation could result from an ill-informed choice.

Durability of foundations can be achieved by selecting and using the appropriate grade of concrete, and correctly constructing the foundation excavation. This will involve control of the water/cement ratio, too much water leading to the likelihood of the concrete being weak. Voids would be formed which would allow the concrete to be compressed under load. Voids could also be left if the concrete is not vibrated to remove pockets of air trapped while placing it.

Foundation concrete and other materials used in the sub-structure construction must be able to carry the required loads without failure. This will require the use of selected graded aggregate. Likewise where the ground contains materials such as sulphates, appropriate sulphate-resisting cement should be used.

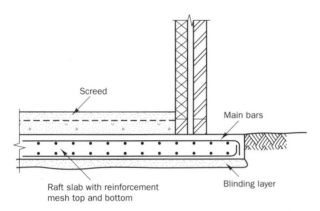

Figure 3.25 Thick reinforced concrete raft

Self-assessment tasks

1. What is the purpose of a foundation?
2. Why are different types of foundations needed?
3. What causes clay soils to move?
4. Explain why the bulb of pressure is useful when selecting a foundation.
5. Describe the factors that could reduce the durability of a foundation.

Walls

Building Regulations require that the walls of a building will be constructed so that any imposed loads are carried safely. This means that the load-bearing walls of a non-framed building must carry:

- the structural dead load
- live loads due to the building being occupied
- the wind loads

Walls transmit the building loads to the foundations. Below ground level it is good practice to use either **engineering brick** or **trench blocks**. Above ground selected facing bricks and thermal blocks are used. The design of the building will determine whether the blocks will be load bearing.

Walls are subject to lateral and vertical loads. Vertical loads will place the wall under compressive stresses and so resistance to **crushing** is required. This can be prevented by:

- **bonding** of the brickwork (Fig. 3.26)
- choosing dense bricks or blocks
- using mortar of equivalent strength to the brickwork

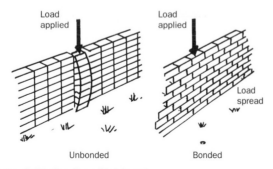

Figure 3.26 Bonding of brickwork

As the height of a wall increases failure is more likely to happen by **buckling** (Fig. 3.27). This is due to the lack of stiffness in the wall which will allow bending to occur. The greater the height of the wall the more the relationship of its height to its thickness becomes important. This relationship is known as the **slenderness ratio**. As the ratio increases the load-carrying capacity of the wall and its stability decreases.

Increasing the stability of the wall by making it thicker is not economic in terms of the use of materials, and in addition it will bring extra loading to the ground. The Approved Documents of the Building Regulations give guidance on the minimum thickness for leaves of cavity walls. There are however two methods which can provide lateral stability and are used to support the wall either at the end or at intermediate points (Fig. 3.28). These are:

- **piers**
- **buttresses**

Piers are columns of masonry which increase the thickness of the wall. They may also support the ends of beams. Buttresses are built at right angles to the wall and may be formed by internal walls or returning external walls, such as in a house or bungalow. Piers are often used in walls forming warehouses and factories.

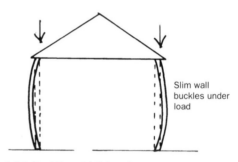

Figure 3.27 Buckling of brickwork

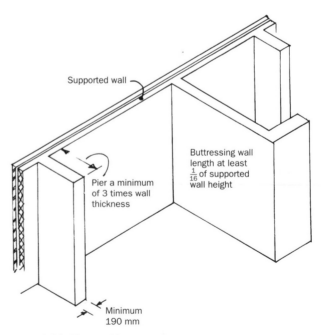

Figure 3.28 Piers and buttresses

Floors

Floors are classed as solid or suspended. Solid floors are those which gain their support from the ground, and suspended are those floors which span between supports. Suspended floors may be at ground level or at intermediate levels within a building, for example in an office block or residential home.

Floors may be constructed from:

- timber and sheet material as the decking
- reinforced concrete
- concrete beams and infill blocks with a screed
- precast concrete floor panels

Whatever the type of floor construction used the ability to carry loads and resist excessive deflection is always required. In addition other properties such as fire, sound and thermal insulation may be needed.

The stability of a **suspended timber floor** will depend upon:

- its span
- the load/s
- depth of joists and their thickness
- spacing of the joists
- type and thickness of the covering forming the floor surface
- use and spacing of strutting
- timber species

Tables A1 and A2 in the Approved Document in the Building Regulations give guidance on acceptable spans for joists. These take into account the depth of the joist, which may have its **effective depth** reduced by notches and holes provided for building services if they are not within the restricted limits.

When the joist span exceeds 2.5 metres it is necessary to provide **strutting** to restrict movement due to twisting and vibration which would damage ceiling finishes. Herringbone or solid strutting can be used (Fig. 3.29).

Where external walls run parallel to internal floor joists **lateral support** can be provided. The support is in the form of

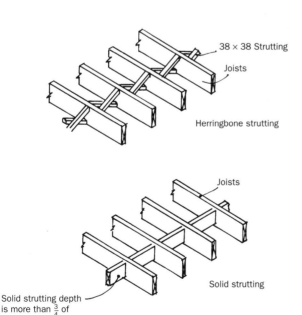

Figure 3.29 Strutting of a timber floor

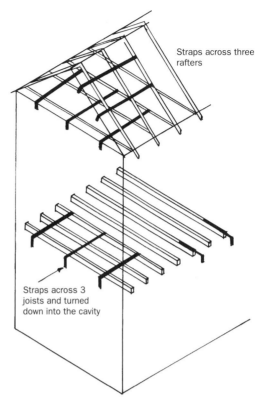

Figure 3.30 Lateral restraint to a wall

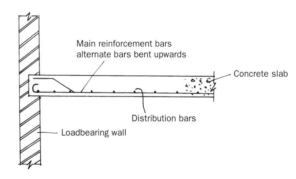

Figure 3.31 Reinforced concrete floor

galvanised mild steel straps of a minimum cross-section of 30 mm × 5 mm fixed across three joists at not less than 2 metres centres (Fig. 3.30). It is usually provided to properties not more than two storeys in height.

Concrete floors will require reinforcement in the tensile area where maximum bending takes place (Fig. 3.31). A floor which is supported on two edges is described as **one-way spanning** and one which is supported on all four edges is **two-way spanning**. The depth of the floor is influenced by the span and loading together with the strength of the concrete. If the floor is built into the surrounding structure it may also provide lateral restraint to the walls, or concrete columns.

Roofs

Common roof forms are flat and pitched. Both must carry:

- their own self weight
- wind load
- snow load

- applied roof coverings
- other imposed loads

Flat roofs are constructed in a similar manner to timber suspended floors and can also afford **lateral restraint** to supporting walls. The construction of a pitched roof is dictated by the shape of the building and the span between supports. This will apply to roofs constructed from timber and steel.

Where there is no internal support the pitched roof is in the form of a triangular truss, which is then subdivided into further triangulated shapes (Fig. 3.32). The structural load-carrying ability will then depend upon the material, shape and size, and the type of joints used to connect the various truss members. These joints give the trusses some of their stiffness. **Timber trusses** are joined using:

- metal truss plates
- bolted connections

Members of **steel roof trusses** are joined together by using:

- bolted connections
- welded connections

When used to form the roof the trusses are spaced at regular intervals. Timber trusses will require **wind bracing** in order to resist side sway collapse. Additional bracing will also provide **lateral restraint** to prevent the slender timber members from deflecting sideways even though the timber will have been stress graded. The bracing also joins the trusses together so the roof acts as one rigid component. Galvanised mild steel straps can also be used to provide lateral restraint to brick gable walls in a similar manner to floor joists (Fig. 3.30).

Steel roof trusses are to be found on factory and warehouse buildings. As with timber trusses a triangulated framework is formed which are inter-connected by purlins which are used to support the roof covering. While this adds some stability to the roof additional wind bracing is often necessary (Fig. 3.33).

Steel may also be used to support the decking and covering of a flat roof. Different forms of beams may be used which include:

- British Standard beams
- braced girders
- castellated beams

Frames

Frame buildings will require effective **lateral bracing** depending upon the stiffness of the joints between the horizontal and vertical components. It is easy to make very

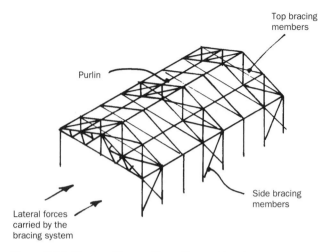

Figure 3.33 Steel roof truss with purlins and bracing

rigid joints when they are formed in concrete, but more difficult when they are formed in steel. When the jointing method cannot provide adequate support the frame can be made more rigid by using:

- **diagonal bracing** (Fig. 3.34a)
- **shear walls** (Fig. 3.34b)

Shear walls are able to prevent the building from collapsing sideways due to the inability of the cladding and roof covering to resist or carry structural loads. Shear walls are often incorporated by designers into the fabric of the building as brick (Fig. 3.35) or concrete panels or internal partition walls. Protected shafts such as staircases and lift shafts are also used to provide additional stability.

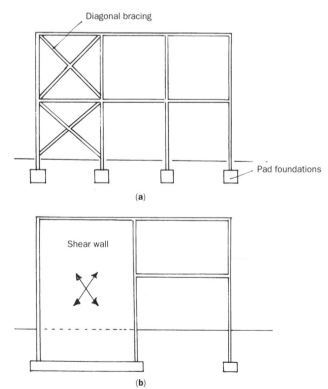

Figure 3.34 Frame bracing: (a) steel frame with diagonal bracing; (b) structure braced by a shear wall

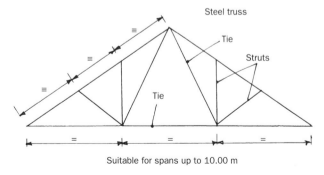

Suitable for spans up to 10.00 m

Figure 3.32 Triangulated roof frame

Figure 3.35 Brick panel walls in a concrete frame building

Self-assessment tasks

1. Identify the primary elements of a building structure and the materials they may be manufactured from.
2. List the loads that structural elements of buildings have to carry and describe the stresses that these loads may create in those elements.
3. Describe five factors that will influence the choice of materials for a skeletal frame.
4. Explain how the stability of a strip foundation and short bored pile foundations is achieved.
5. Describe the effect that the slenderness ratio has on buckling of a wall, and how buckling may be prevented.
6. Identify five factors that could affect the stability of a timber suspended floor.
7. With the aid of a simple sketch explain how tensile stresses are reduced in a concrete floor slab.
8. Describe how wind load is transmitted from a roof covering to the foundation of a framed building.
9. Prepare a presentation to show how a domestic building carries dead and imposed loads safely to the ground.

3.2 Construction of low-rise buildings

Construction sites are one of the most dangerous working environments. This in part can be attributed to the ever changing situation as work progresses and also to the lack of training and safety awareness of the operatives. Falls from scaffolding or into excavations contribute to a high percentage of accidents. It is therefore important that safe and adequate access is provided to allow operations to be achieved without accident or injury. The following legislation places requirements on the employer to provide safe working practices and procedures:

- Health and Safety at Work etc. Act 1974
- The Construction (Lifting Operations) Regulations 1961
- The Management of Health and Safety at Work Regulations 1992
- The Provision and Use of Work Equipment Regulations 1992
- The Construction (Health, Safety and Welfare Regulations 1996
- The Workplace (Health, Safety and Welfare) Regulations 1992
- The Construction (Design and Management) Regulations 1994

As the building structure rises operatives will need a safe working platform. This may be provided by:

- **scaffolding**
- **towers**

These in turn will need to have proper means of access; normally this will be provided by **ladders**. Building materials and people can be transported by means of a **hoist**.

Ladders

Where work cannot be carried out safely from the ground ladders may be used to access the work, but only if light work is to be carried out. British Standards 1129 and 2037 cover different types of ladders. Duty ratings are used to show their suitability for particular uses. The ratings are:

- Class 1   Industrial – heavy duty
- Class 2   Light trades – medium duty
- Class 3   Domestic – light duty

Ladders may be made from timber or aluminium. They must not be painted, or used if defective in any way. Metal ladders should not be used near electrical equipment or supplies. Ladder inspections should be carried out which should record:

- date of inspection
- defects
- repairs
- issue and return

When using ladders the following should be observed:

- constructed of a sound, adequate material
- they should be placed on a firm level base

- adequate lashing either near the top or bottom
- the top of the ladder must extend beyond a landing place on a scaffold by 1.0 m
- ladders under 5.0 m may be 'footed' if they cannot be fixed or lashed
- the ladder angle should be as near as possible to 75° or one rung out for every four up (Fig. 3.36)

Ladders should not be used if there is evidence of:

- damaged stiles
- missing or broken rungs
- mud or grease on the rungs
- rotted timber or defective fittings
- sagging, warping or distorting

Common forms of ladder are:

- stepladders
- pole ladders up to 10 m long
- extending ladders up to 20 m long
- roof ladders

Scaffolding

The use of scaffolding allows a safe working platform to be constructed and modified as a building progresses upwards. In

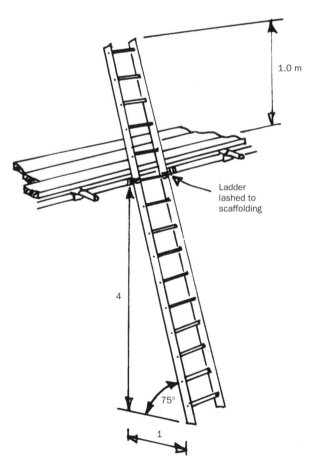

Figure 3.36 Use of a ladder

addition safe access and egress to the scaffold should be maintained. Only competent persons should erect or modify scaffolding and carry out the statutory inspections. Various types of scaffold can be constructed according to the location and use. Table 3.8 describes scaffolding components.

Table 3.8 Scaffolding components and use

Component	Application
Sole plate	Used to spread the load from the standard
Standard	Vertical metal poles used to support transoms, putlogs and ledgers
Transom	A horizontal member fixed to the standards supporting putlogs or transoms
Putlog	A horizontal member used on a putlog scaffold to support scaffold boards, one end being carried in the joints of brick wall
Ledger	A horizontal member used on an independent scaffold which carries the scaffold boards
Bracing	Used to stabilise the scaffold. May be longitudinal across the face of the scaffold and standards, or transverse bracing the ledgers
Ties	Used to stabilise and tie the scaffold to a building structure through openings in the building
Scaffold boards	Normally in timber and used to form the working platform of various widths. Ends are protected by steel bands
Guard-rail	Used as a safety barrier to prevent people falling from the scaffolding
Toe board	Used with guard rails to prevent the accidental movement of tools and materials over the edge of the scaffold
Nets	Used to prevent objects falling from the scaffold, may be flexible or rigid

Putlog scaffolding

The putlog scaffold consists of a single row of standards and ledgers which support putlogs with a spade end which are built into a supporting wall (Fig. 3.37). The putlogs in turn support the working platform. The working platform is then guarded by toe boards and guard rails to which are fixed protective mesh. Access to the work area is gained by a lashed ladder extending beyond the working platform by 1.0 m. This scaffold is often used in conjunction with new brick built structures.

Independent scaffolding

This scaffold has two rows of standards which support ledgers and transoms to form the working platform (Fig. 3.38). The platform is guarded in a similar manner to the putlog scaffold. Despite its name the scaffold is not independent from the building. Tying into the building is necessary which is achieved by using ties through openings called **through ties** or into reveals by using **reveal ties**. Additional stability of the scaffold is achieved by using longitudinal and transverse bracing. This scaffold is used with steel or concrete frame buildings or in situations where maintenance or alteration work is taking place on an existing building. Traditional and patent scaffolding systems are available.

Figure 3.37 Putlog scaffolding with mesh closing the gap between toe board and guard rail

Figure 3.38 Independent scaffolding

Working platforms

Working platforms and decking should be close boarded to their full width. Care should be taken to ensure that tripping cannot occur. If boards overlap each other there should be bevelled pieces to reduce tripping over the ends of boards.

The scaffold platform should be of an adequate size to match the work to be carried out on the scaffold. Widths can vary from two to seven boards in width. Five boards wide is a common size for bricklaying operations. Scaffold boards should overlap putlogs or transoms by a minimum of 50 mm. Maximum overhang varies with board thickness from 150 mm to 250 mm.

Towers

There are occasions when local access is required to part of a building for inspection or general maintenance. Tower scaffolds allow this. They are either mobile or static and most commonly in proprietary form. The height of the towers should not exceed three times their width. Raking bracing may be necessary to ensure further stability.

Hoists

Hoists are used on sites to:

- move goods and materials
- move passengers

They enable the efficient and rapid movement of people or goods and contribute to maintaining productive environments on the site. The following should be observed when using hoists:

- they must be erected by a competent person
- ground conditions should give adequate support
- forces caused by inclement weather
- hoist operators should be trained
- statutory regulations on inspections should be complied with

Hoists may be stand-alone or integral with a scaffolding system. Whatever the situation the hoist ways must be protected at ground level and at all the access points. Guard gates should be at least 2.0 m high and kept closed at all times except to allow the movement of goods or people. Operation of the hoist should only be possible from one position at a time, with a clear view of all the levels.

Self-assessment tasks

1. Describe how ladders should be secured when used to provide access to a scaffold platform 6.00 m above the ground.
2. Why is it necessary to inspect ladders and record details of the inspection?
3. Explain briefly the differences between a putlog scaffold and an independent scaffold.
4. How is the stability of scaffolding achieved?
5. What steps should be taken to ensure that the working platform is safe to work from?

Forms of sub-structure

The **sub-structure** is that part of the structure situated below the damp-proof course of a building. Its purpose is to transfer the dead and imposed loads of the superstructure to the ground. An essential part of the sub-structure is the **foundation**.

Foundations

Their function is to carry the loads of the building and distribute them over the ground in such a way that movement of the building is minimal. Different types of foundations are used according to:

- building load
- structural form of the building
- type of sub-soil

Strip foundations are used with load-bearing walls (Fig. 3.21) and in ground conditions where the load from the foundation is able to be carried safely within 3 metres of the surface. Walls are placed centrally on the foundation to ensure that there is uniform distribution of stresses to the soil beneath.

Mass concrete is used to form the foundation and should be as specified in BS 5328: Part 2 for Grade ST1 concrete. Sulphate-resisting cement may be required where there are aggressive chemicals in the ground. For simple strip foundations punching through of the wall is resisted by making the thickness of the concrete at least equal to the projection from the wall (Fig. 3.21). Alternatively it may be reinforced as shown in Fig. 3.24.

Common site practice today is to excavate and fill the whole of the trench with concrete to form what is known as **deep strip** or **trench fill** foundations (Fig. 3.22). These are also known as **mass concrete** foundations. The depth of concrete is normally sufficient to resist movement and maintain stability.

Short bored piles also consist of mass concrete for the bulk of their form and are reinforced towards the top to link with the reinforced ground beams (Fig. 3.23). The ground beams act as a reinforced strip foundation to carry brickwork.

An alternative to deep strip foundations is the **reinforced strip foundation** (Fig. 3.24). Steel reinforcement is placed across the width of the foundation so that the tensile bending stresses are resisted and punching through prevented.

Reinforcement is also used for **concrete raft foundations** (Fig. 3.25). Depending upon the ground conditions the whole slab or edge and intermediate beams may be reinforced.

In areas where the soil investigation has shown that the ground-bearing capacity is weak it will be necessary to either remove the weaker soil and replace with well **compacted selected fill**, or use reinforcement in the strip foundation to bridge the weak area.

Function of superstructure elements

The superstructure is normally taken to be that part of the building above the ground or the damp-proof course. Its purpose is to carry the loads imposed on the building as dead and live loads safely to the substructure which then transmits them to the ground.

Elements of the superstructure include:

- walls
- floors
- roofs
- frames

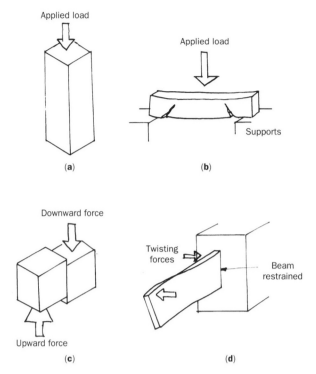

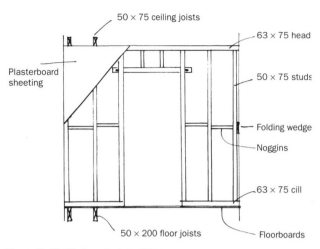

Figure 3.40 Timber stud partition

Figure 3.39 Types of structural stress: (a) compressive; (b) tensile; (c) shear; (d) torsion

All of these components must remain **stable** under load. In order to understand how stability should be achieved the structural behaviour of the components should be known. This behaviour will be influenced by:

- structural loads and external forces
- materials used
- size and position of the components

These factors will then influence the type of structural stress which may be:

- compressive (Fig. 3.39a)
- tensile (Fig. 3.39b)
- shear (Fig. 3.39c)
- torsion (Fig. 3.39d)

The result of these stresses may be:

- buckling in vertical members – walls, columns
- bending in horizontal members – beams, floors
- failure of members – ties in roofs, beams
- twisting of members – columns, beams, girders

Walls

Wall stability has been discussed previously. However the influence of the following on the wall should not be forgotten:

- type of materials used – brick, concrete, mortar
- construction of the wall – bonding, cavity ties
- design of the wall – slenderness ratio
- wall stiffening – piers, buttresses, lateral restraint ties
- type of load

Timber stud partitions may be used to form internal walls. Normally they are non-load bearing but can be designed to carry structural loads. The construction is shown in Fig. 3.40.

Different forms of sheeting may be applied to the studs but where fire resistance is required plasterboard and skim are used. The middle of the partition can be filled with insulation which will improve its thermal and acoustic properties.

Load transmission in walls occurs through the bonding of the brickwork (Fig. 3.26). The load that can be carried is determined by:

- density of the brick
- strength of the mortar
- type of brick bond used

Brick compressive strengths range from $5 \, \text{N/mm}^2$ to $7 \, \text{N/mm}^2$ and blocks from $2.8 \, \text{N/mm}^2$ to $15 \, \text{N/mm}^2$. Mortar proportions should be as given in BS 5268 Part 1 or $1 : 1 : 6$ Portland cement, lime and sand.

Where cavity walls are used the cavity must be at least 50 mm wide and the separate leaves supported by **cavity ties** at 900 mm spacing horizontally and 450 mm spacing vertically to assist with load transmission and prevent the leaves from buckling.

Where doors and windows are placed in walls the structural load must not be carried by these components. Two methods (Fig. 3.41) may be used to divert the load:

- **lintels**
- **arches**

Lintels are a form of beam which spans across an opening and provides support for the wall above. Lintels may be formed from concrete or steel beam or pressed steel (Fig. 3.42).

Arches span openings by using small units of brick or stone and are laid to a profile. Common profiles used are segmental and flat arches (Fig. 3.43).

Durability of walls is related to the degree of weather exposure they are subjected to in a particular location. Again choice of brick and mortar is important in order to minimise the amount of moisture entering the brickwork which may then in turn become subject to frost action in winter.

Damp-proof courses (dpc) are used to prevent the passage of moisture from the ground up a brick wall (Fig. 3.21). The dpc is placed at least 150 mm above ground level and in both leaves of a cavity wall. Damp-proof courses are also placed under copings on parapet walls and around door and window openings, and chimneys.

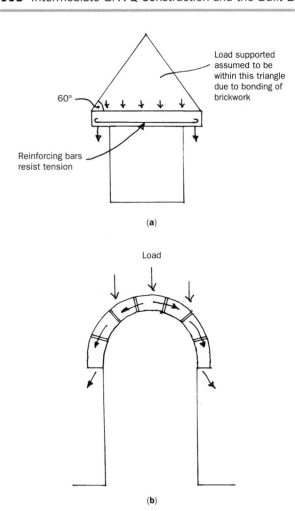

Figure 3.41 Brickwork load transmission: (a) lintel; (b) arch

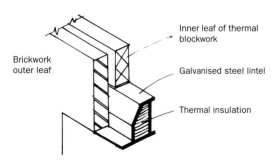

Figure 3.42 Steel lintel in a cavity wall

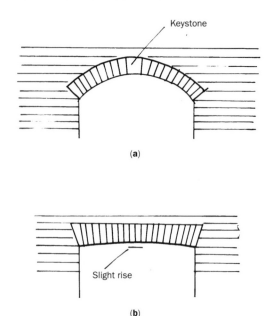

(a)

(b)

Figure 3.43 Brick arch in a wall: (a) segmented; (b) flat

Fire resistance of walls involves two aspects:

* surface spread of flame
* inherent fire resistance of the wall

Ideally all materials should not assist with the spread of flame across their surface. Bricks have high non-combustible properties and are therefore Class 0 for surface spread of flame, meaning no spread will occur.

A wall's inherent fire resistance can fail when:

* the wall collapses under fire
* fire breaks through part of the wall
* heat insulation properties of the wall are destroyed

No material or form of construction is totally fireproof and therefore periods of fire resistance are quoted for specific building uses in specific situations. Minimum periods vary for load-bearing walls but range from 30 minutes to 120 minutes.

Internal walls may be required to be fire resistant; typical situations are:

* along designated fire escape routes
* protected shafts – lifts, staircases
* compartment walls between different use of spaces
* load-bearing walls

Thermal transmission through walls can be quite high because of the density of the materials used. The insulation value of a wall is measured by its **U value**. The lower the U value the better the insulation value of the wall. Currently a U value of $0.45 \, \text{W/m}^2\text{K}$ is required for external walls.

This can be achieved by:

* careful selection of materials
* use of a cavity wall
* insulation in the cavity
* using thermal blocks on the inner leaf
* dry lining the internal leaf

Sound resistance of external walls does not normally present a problem unless a building is situated next to a major road, railway, airport or noisy industrial area. The use of dense materials, cavity wall with insulation, and double glazing will assist. In severe locations it may be necessary for air conditioning to be installed to prevent noise through open windows.

Within the building, walls which separate semi-detached or terraced houses must be capable of keeping noise down between adjoining properties. Requirements vary according to the form of construction used so the Approved Document E to the Building Regulations should be consulted.

In simple situations where the mass of the wall is used to reduce sound, the wall including the finish should have a density of $375 \, \text{kg/m}^2$, and concrete blocks or in-situ concrete a density of $415 \, \text{kg/m}^2$.

Floors

Floors should be strong enough in order that they can:

- carry the required loads
- reduce deflection to a minimum
- afford support to other parts of the structure

Timber floors

Some aspects of the **stability of timber floors** have been previously discussed in section 3.1. In addition to those points made there suspended timber ground floors are able to gain intermediate support from sleeper walls which enable a reduction in the joist span and the size of the joist (Fig. 3.44). Further stability is provided by the joists being built into the external and sleeper walls thus reducing the need for strutting.

Suspended timber upper floors have to span from wall to wall and therefore they are deeper in depth to limit deflection and will require strutting to reduce twisting and sideways movement of the joists. The joists may be built into the walls or suspended from joist hangers. The stability of the floor surface will depend on the spacing of the joists and the board or sheet material used. 18 mm is considered to be a minimum thickness for boards and sheets.

Often timber floors are required to support a timber stud partition. This will bring an increased dead load in a particular area of the floor. In order to carry this load adequately double joists are used, often bolted together (Fig. 3.45a). Where the partition runs across the joists a wall plate is used (Fig. 3.45b).

Upper floors often have a staircase to accommodate. This will involve shortening some of the joists to create a **stairwell**. Figure 3.46 shows how the main bridging joists are cut and supported by a trimmer joist which in turn is supported by trimming joists.

Deterioration of ground floors can occur which in turn will affect their **durability**. To prevent this damp-proof courses must be laid under all the joists where the wall is in direct contact with the ground. Ventilation of the air space in order to prevent dry rot is achieved by using **air bricks** and honeycomb sleeper walls (Fig. 3.44). Where joist ends are built into the wall a timber preservative should be used.

Ground floors are required to meet **thermal insulation**

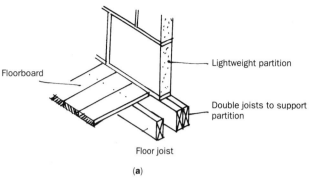

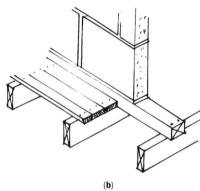

Figure 3.45 Floor support to a lightweight partition: (a) double joist; (b) timber wall plate

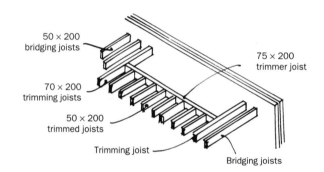

Figure 3.46 Stair well in a timber suspended floor

requirements of the Approved Document L of the Building Regulations. The student is advised to refer to these for specific situations. Usually a U value of $0.35 \, \text{W/m}^2 \, \text{K}$ to $0.45 \, \text{W/m}^2 \, \text{K}$ is required.

Sound insulation is not normally a requirement with timber suspended floors unless a property has multiple occupancy, say flats. In this case upgrading of the floor can be achieved by creating a new ceiling with two layers of plasterboard to which is added layers of mineral wool and a floating insulation layer (Fig. 3.47).

Fire resistance for timber floors requires that they are able to:

- continue to carry a load for a stated time
- prevent fire from breaking through
- resist excessive heat transfer

This may be achieved for example by using 12.5 mm plasterboard nailed to joists at 600 mm centres with 40 mm galvanised nails at 150 mm centres. The plasterboard must be fixed to every joist and the heading joints nailed to timber noggins.

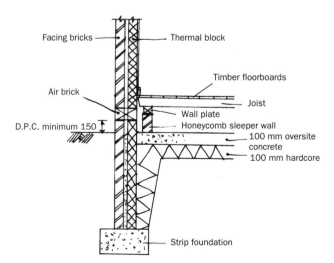

Figure 3.44 Suspended timber ground floor

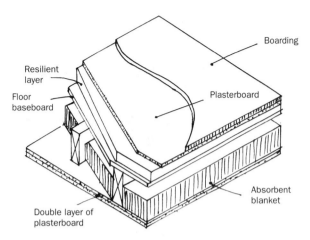

Figure 3.47 Sound insulation to suspended floors

Concrete floors

Concrete is weak in tension and strong in compression. When used to form a ground floor slab it is supported from the ground which reduces to a minimum the tensile stresses. In order that the floor remains stable it is cast on top of a bed of hardcore. This hardcore has replaced the topsoil which is a compressible material. The slab may have a damp proof membrane (dpm) placed between it and the hardcore. It is essential this dpm is taken up to meet the dpc in the wall to prevent moisture bypassing the dpc (Fig. 3.48).

When used in a suspended floor concrete is placed in tensile stress. In order for a suspended floor to be **stable** and **transmit loads** satisfactorily mild steel reinforcement is used. This is placed in the tensile area of the slab (Fig. 3.31) to reduce bending and hence deflection of the floor slab. Reinforcement is necessary to reduce shear stresses where the slab is supported.

Simply supported slabs are one way spanning. Where heavy loads are to be carried the floor is reinforced in both directions and becomes two way spanning. Other forms of concrete slab are used. These can reduce the volume of concrete being used. Typical ones are:

* trough floors
* hollow pot

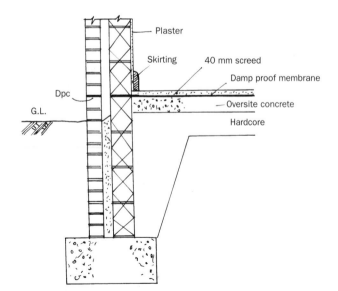

Figure 3.48 Section through a concrete floor

Fire resistance of concrete slabs is good initially up to 120 °C but thereafter there is a progressive loss of strength. An important consideration is that the steel reinforcement must retain its concrete cover as long as possible. Once this cover is lost failure of the slab will be quick. The minimum cover of concrete is specified according to the anticipated fire load on the structure. The minimum depth of cover is 25 mm.

Concrete is not a good **thermal insulator**. In order to satisfy its load-carrying requirements its density is most important. Unfortunately dense materials conduct heat at a faster rate than those with voids in them. Concrete does however have good **resistance to airborne sound** and poorer qualities with respect to structure borne sound.

Roofs

There are many shapes of roofs but most fall into the two general categories of flat and pitched. A flat roof has its external face at less than 10° and a pitched roof over 10° to the horizontal.

Flat roofs

Flat roofs are very similar to timber floors in their construction. They have roof joists which support a decking which in turn supports a water resistant roof covering. They also span between supports in a similar way to a suspended timber floor (Fig. 3.49). Openings for items such as roof lights can be trimmed in a manner similar to stairwells.

Stability of the roof is achieved by using:

* roof joists having an adequate cross-sectional size
* timber which is stress graded
* strutting to stiffen the joists laterally
* galvanised mild steel anchor straps

Concrete may also be used to form a flat roof slab. It is reinforced in a similar manner to a suspended floor slab. Unless the concrete is of a very high quality it will need a weatherproofing membrane on its outer surface.

Timber flat roofs are not naturally **fire-resisting** constructions though the application of plasterboard and skim to the underside of the joists will provide at least half an hour's fire resistance. Where the flat roof passes over the party wall between dwellings fire stopping will be necessary to prevent fire spread between dwellings. A typical example is shown in Fig. 3.50.

Roof finishes should ideally have a low spread of flame but a flat roof with a built up felt finish with 12.5 mm bitumen bedded stone chippings will be acceptable. Concrete will possess sufficient fire resistance for most purposes.

Airborne **sound resistance** of a concrete flat roof is excellent, any weak points being where there may be roof openings. Timber roofs are not normally required to provide high levels of sound resistance.

Thermal transmission through flat roofs can be enhanced by the addition of thermal insulation. In the case of the timber flat roof the roof can either be a warm deck roof or cold deck roof construction. **Warm deck roofs** can have the insulation placed under the surface of the waterproofing layer, known as **sandwich construction** or on top of the waterproof layer which is known as **inverted construction**. **Cold deck roofs** have the insulation placed in the roof void between the joists (Fig. 3.51).

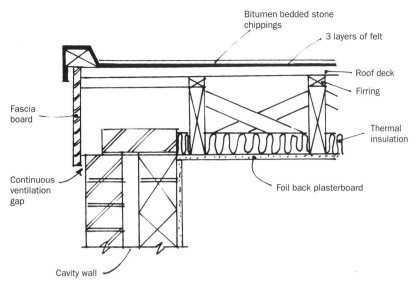

Figure 3.49 Flat roof construction

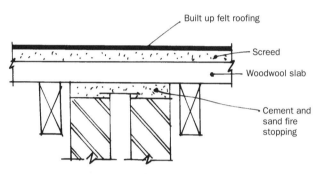

Figure 3.50 Fire stopping in flat roofs

All three decks have disadvantages but the cold deck is considered to be the least desirable because of the risk of condensation and the variability of the wind speed which ventilates the roof.

Pitched roofs

Some factors which affect roof construction and structural stability have been noted in section 3.1. Whether the roof is made up from steel or timber it is still required to carry and transmit all the loads placed upon it safely.

All pitched roofs rely on triangulated frameworks for their **stability**. Traditional timber roofs (Fig. 3.52) consist of the following members:

- rafters
- purlins
- bracing
- collars
- hangers
- ceiling joists
- binders
- ridge board
- wall plates

It can be seen from Fig. 3.52 that there are several triangulated frames which maintain the **stability** of the roof construction. The whole roof is in effect a series of parallel

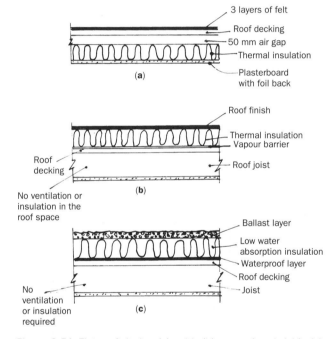

Figure 3.51 Flat roof decks: (a) cold; (b) warm (sandwich); (c) warm (inverted)

frames joined together by the purlins, binders, ridge board and bracing.

The trussed rafter (Fig. 3.53) is the modern equivalent of the traditional cut roof. Every position has a braced truss which has led to timber economies and increased strength characteristics. These are spaced at 600 mm centres. Further **stability** is provided by wind bracing and strapping.

The load of the roof covering is carried firstly by the tile battens to the rafters and then the **timber roof transmits the loads** to a wall plate. This is bedded in mortar on top of the inner leaf of the wall and spreads the load evenly along the wall. The stability of the wall plate is assisted by galvanised straps fastened to the wall.

Steel roof trusses are constructed in a similar manner using

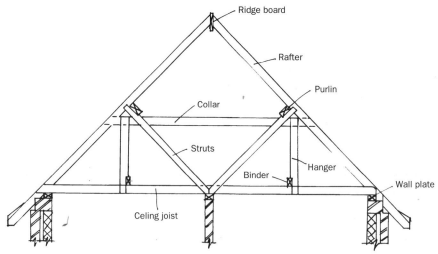

Figure 3.52 Traditional timber roof

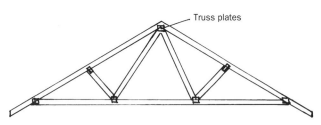

Figure 3.53 Trussed rafter

angle section members to form rafters, struts, purlins and ties. These are connected together with bolted or welded gusset plates. The **load transmission** of the roof load is via the purlins to the trusses which are spaced at 3.00 m to 4.5 m centres.

Steel roof trusses normally sit bolted onto mild steel plates on the tops of columns. If the building structure is brickwork then piers will have been built with a concrete pad stone on the top to carry the load down to the foundation.

Fire resistance of pitched roofs will rely mainly on applied finishes to the roof surface and to the underside or soffit of the roof. Timber will tend to smoulder and char therefore keeping its structural strength for a long period of time.

Steelwork will remain stable until the fire reaches the steelwork and a temperature of 250 °C is reached, beyond which there is a rapid loss of strength. Fire protection can be provided by applied cladding, intumescent paint or sprayed concrete. As for flat roofs, fire stopping is required to prevent spread between properties at roof level.

Thermal insulation can be installed to create either a cold or warm roof structure. Principles are the same as those outlined for flat roofs. Profile sheeting for warehouse and factory applications on steel trusses can incorporate a thermal sandwich which will form a warm roof structure. Building Regulations expect mimimum U values to be achieved. Approved Document L gives further guidance.

Roofs are required to prevent moisture reaching the inside of the building. Roof tiles in many materials and forms are used. Profiled PVC, metal, and sheet roof coverings are also available. Gutters and downpipes are used to collect the rainwater and carry it to a safe disposal point away from the building.

Self-assessment tasks

1. Explain how the sound and thermal insulation qualities of a timber stud partition can be improved.
2. Identify three factors that determine the load that can be carried by a brick wall.
3. Make sketches to show two ways of carrying the structural load above a window.
4. How can the thermal resistance of a cavity wall be increased?
5. Explain how a suspended timber floor can safely support a timber stud partition.
6. Why should plasterboard be nogged at every heading joint when seeking to achieve half hour fire resistance from a timber floor?
7. What is the purpose of the dpm in a concrete ground floor, and what precaution should be taken at its edges?
8. Use simple sketches to illustrate the terms warm and cold deck roofs.
9. What purpose do purlins serve in pitched roof construction?
10. Prepare a diagram which demonstrates the performance requirements for a two-storey house.

Function of secondary elements

Secondary elements are those parts of a building which normally do not carry structural load but contribute towards the functional operation of the building. The principal elements involved relate to:

- doors and associated openings
- windows and associated openings
- stairways, landings and balustrades
- building services and fittings
- finishes

Doors

Doors are part of the enclosing and dividing function of a building and are fitted within walls or partitions. They can be classed generally as internal or external doors. Within each of

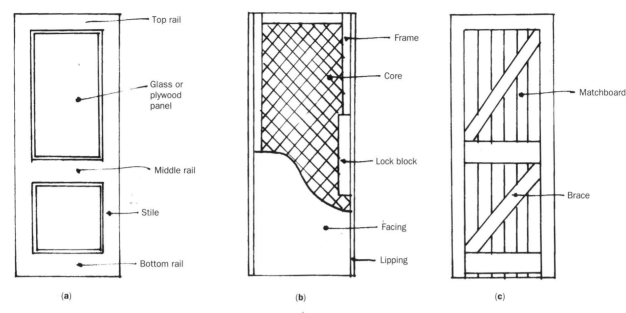

Figure 3.54 Timber doors: (a) panelled; (b) flush; (c) matchboarded

these classes are further types of doors. Doors may be made from:

- timber – hardwood or softwood
- metal – steel or aluminium
- rigid uPVC

Timber is commonly used for external domestic doors though aluminium and uPVC are being used increasingly. Metal doors are often used for shops and offices, industrial and commercial buildings. Stainless steel and bronze may be used for doors in prestigious buildings such as banks and building societies.

Timber doors may be:

- panelled
- flush
- match boarded (Fig. 3.54)

These are available in a range of co-ordinated sizes. Standard sizes available are 1981 mm high × 762 or 838 mm in width. The frames correspond to the dimensions of brickwork courses into which the door will fit.

Aluminium doors may be hinged or sliding. They are made from extruded sections and can accommodate panels and double glazed units. They are supported in hardwood timber frames fully weather-stripped.

Steel casement doors are made from rolled steel sections to British Standard profiles. Glazing units and sheet panels can be provided. Thermal insulation can be enhanced by using double glazing and bonded insulation in panels.

Rigid uPVC doors use hollow extrusions which may need reinforcement with internal metal sections in the hollow cores. Double glazing and weather-stripping are standard features.

Typical performance requirements for doors are shown in Table 3.9

Stability of doors is normally achieved by using a basic framework of members called:

- stiles
- rails

Table 3.9 Performance requirements for doors

Performance	Factors to consider
Access and egress	Who will use the door. Family, old or disabled people, general public, vehicles
Resistance to weather	Controlling draughts and heat loss, preventing water penetration. Fit of the door into the frame, and the frame into the structure
Security	Type, location, and use of the building. Selection of glass and ironmongery
Passage of heat	Draught strips and double glazing
Noise control	Location relative to source and type of noise. Weight of the door
Affording privacy	Reason for privacy, e.g. an office in a shop, and the means for achieving it, e.g. opaque glass
Fire resistance	Locations such as flats, garages, and protected routes
Stability	Minimising deformation due to expansion and contraction and use

- muntins
- bracing
- joints

In addition the firm fixing of the door frame into the structure is important (Fig. 3.55).

Durability of doors is achieved by:

- careful selection of materials
- using watertight joints
- using suitable preservative treatments
- protecting vulnerable parts (using kicking plates)

Safety is assured by using

- good quality locks
- safety glass
- correctly adjusted closing mechanisms
- correct choice and location of ironmongery

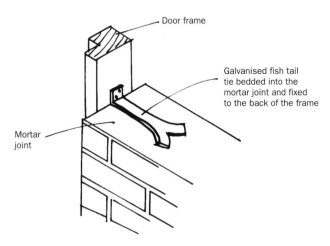

Figure 3.55 Fixing of a door frame to a wall

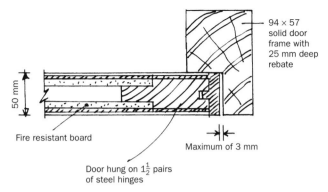

Figure 3.56 Fire door and frame detail

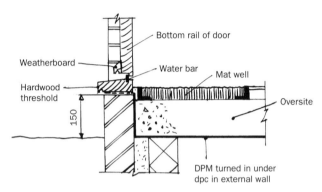

Figure 3.57 Resistance to moisture penetration

Resistance to fire is achieved by:

*   using materials resistant to penetration of flame or hot gases
*   careful design of both door and frame (Fig. 3.56)
*   using Georgian wired glass in panels
*   choice of ironmongery and fittings

Sound resistance is not often required except in special cases, any reduction effect normally being achieved as a by-product of the materials used for the door construction. Specialist doors are available but are often for internal use only. Sound reduction relies on the weight of the door and good seals being achieved between the door and frame.

Resistance to impact will depend upon:

*   the cause of the impact
*   type of door construction
*   materials used
*   material used to cover the external surface of the door

Resistance to moisture penetration will rely on:

*   high quality door construction
*   the fit of the door into its frame (Fig. 3.57)
*   weather-stripping in the rebates
*   good threshold design
*   the design of outward-opening doors

Doors in order to function must be supported in a frame. It is common practice to hang external doors in **door frames** (Fig. 3.58) and internal doors in **door linings** (Fig. 3.59).

Timber door frames are formed in one piece with the moulding and rebate being created by machine. The frame is fixed into the opening making sure that the vertical dpc tucks in the back of the frame to prevent moisture penetration (Fig. 3.60). The component parts of a door frame are the:

*   jambs
*   head
*   threshold
*   rebate

Door linings are lighter frames to which internal doors are hung. They may be made to accommodate the door alone or extend from the floor to the ceiling when they are referred to as storey linings (Fig. 3.61). **Doorsets** are also used which comprise the door and its frame together with the architraves, fan light, side panels, and ironmongery as appropriate.

Self-assessment tasks

1. What performance factors would be considered important for a door between a house and an integral garage?
2. Describe how the stability of a door frame is achieved.
3. Use a simple sketch to show how moisture penetration is prevented at the bottom of an external timber door.
4. Explain using sketches the terms 'door frame' and 'door lining'.
5. What are the advantages of using 'doorsets'?

Windows

Windows are used in buildings to:

*   admit daylight and sunlight
*   allow an external view
*   permit ventilation of a space

In performing these functions the window may also be required to:

*   exclude moisture
*   minimise heat loss
*   reduce sound transmission
*   remain stable
*   be durable for an acceptable period of time
*   withstand fire for a minimum period (Fig. 3.62)

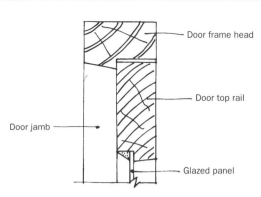

Figure 3.58 Door frame

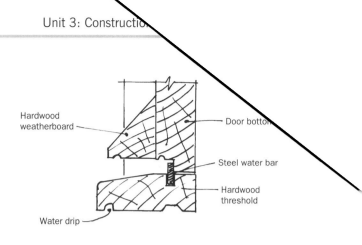

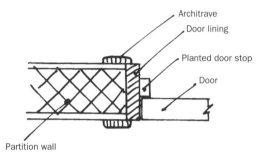

Figure 3.59 Door lining

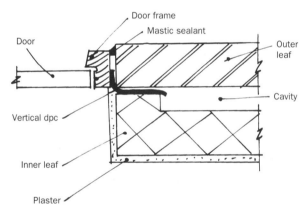

Figure 3.60 Dpc detail to a door frame

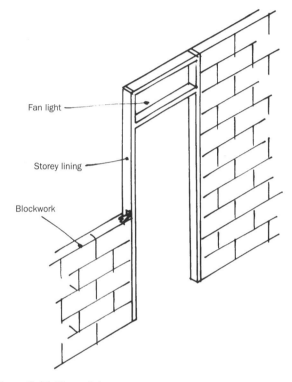

Figure 3.61 Storey lining

Table 3.10 Performance requirements for windows

Performance	Factors to consider
Resistance to weather	Controlling draughts and heat loss, preventing water penetration. Fit of the window sash into the frame, and the frame in to the structure
Security	Type, location, and use of the building. Selection of glass and ironmongery
Passage of heat	Draught strips and double glazing
Noise control	Location relative to source and type of noise. Thickness and spacing of glass
Affording privacy	Reason for privacy, e.g. an office or a bathroom, and the means for achieving it, e.g. opaque glass
Fire resistance	Locations such as flats, and protected routes. Use of wired glass
Stability	Minimising deformation due to expansion and contraction and use
Providing ventilation	Satisfying Building Regulations

Windows can be classified as:

- casement
- sliding (horizontal or vertical)
- pivoted (vertical or horizontal)

Windows may be manufactured from:

- timber – hardwood or softwood
- metal – steel or aluminium
- rigid uPVC

Stability of windows is normally achieved by using a basic **framework** of members (Fig. 3.63) called:

- cill – a horizontal member at the bottom of the frame
- head – a horizontal member at the top of the frame
- transom – an intermediate horizontal member
- mullion – an intermediate vertical member
- joints – usually mortise and tenon to connect the above members together

In order to allow ventilation to take place opening parts of the window called **sashes** or **casements** are used. These are made from:

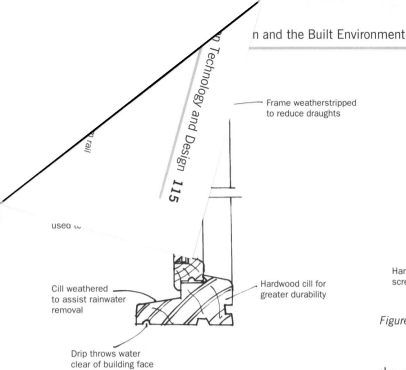

Figure 3.62 Window section

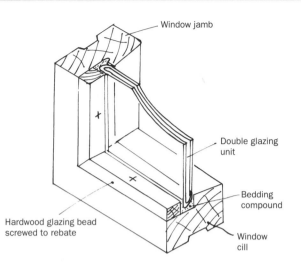

Figure 3.64 Double glazing

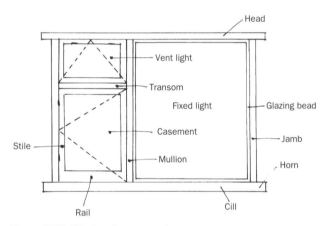

Figure 3.63 Window frame members

- top and bottom rails
- stiles
- glazing bars

In addition the firm fixing of the window frame into the structure is important. This is achieved in a similar way to the door frame by using frame cramps.

Durability of windows is achieved by:

- careful selection of materials
- using watertight joints
- using suitable preservative treatments
- protecting vulnerable parts (cills, bottom rails, transoms)

Security and safety is assured by using

- good quality locks
- safety glass
- correctly adjusted closing mechanisms
- correct choice and location of ironmongery

Sound resistance is not often required except in special cases. Some reduction is achieved as a by-product of double glazing for reducing thermal transmission. Sound reduction relies on the weight of the glass and distance between the panes and good seals being achieved between the sash and frame. An air gap of 100 mm upwards is usual to improve sound insulation properties.

Thermal transmission is controlled by using double glazing with two panes of glass 10 to 20 mm apart (Fig. 3.64). The sheets of glass are hermetically sealed at the edges using a metal spacer and bedded in a non-setting compound. Flexible sealing strips are used to prevent air leakage between the sash and frame. Windows made from uPVC often have the cavity of the extruded section filled with polyurethane foam to reduce thermal transmission and condensation.

Resistance to moisture penetration will rely on:

- high quality window construction
- the fit of sashes into the frame, and the frame into the opening
- weather-stripping in the rebates
- good cill and transom design
- careful design of opening sashes

Building Regulations and Approved Documents require habitable rooms to be ventilated and in particular:

- kitchens
- bathrooms
- toilets
- living rooms
- bedrooms

Two forms of ventilation are required:

- rapid
- background

Rapid ventilation is usually provided by either a door or window. Windows should have at least part of them 1.75 m above floor level. The openable part of the window should have an area of at least one-twentieth of the floor area. Background ventilation should be not less than 8000 mm^2. Kitchens and bathrooms will require half of this figure but must have a form of extract ventilation. Further specific requirements exist for utility rooms and sanitary accommodation.

Stairs

Staircases are used to enable people or goods to be moved from one level to another either within or outside a building. A further function is to provide an effective means of escape in case of fire.

Stairs may be individually designed and constructed or purchased from a factory-made range. Materials used to manufacture stairs are:

- timber – softwood or hardwood
- concrete
- steel
- a combination of the above materials

Table 3.11 Terminology associated with stairs

Term	Definition
Flight	A series of steps between a landing
Landing	A floor area at the top of, or between flights of stairs
Tread	Horizontal surface of a step
Winder	A tapered tread used to allow a stair to change direction while continuing to climb
Riser	Vertical surface of a step
Nosing	The part of the tread which overhangs the riser
Going	The horizontal distance between two consecutive nosings
Rise	The vertical distance between two consecutive treads
Pitch line	An imaginary line joining the nosings of a flight of stairs
Pitch	The angle formed by the pitch line and the horizontal
Headroom	The vertical distance between the pitch line and the ceiling above it
String	A side member of the stairs which supports and carries the steps
Handrail	A support fixed parallel to the pitch line at the sides of a stair
Balusters	Vertical members supporting the handrail
Newel post	A post which can support the string and the handrail to a stair

Stairs are made to meet the requirements of the Building Regulations. The Approved Documents require compliance with:

- pitch angle
- headroom
- handrail height
- baluster spacing
- tread and riser sizes
- guarding to landings
- landing size
- access for the disabled
- fire resistance periods

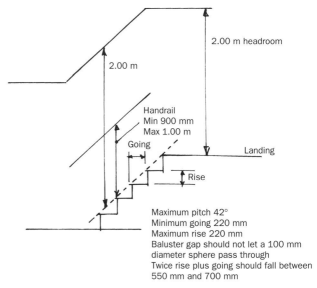

2.00 m headroom

2.00 m

Handrail
Min 900 mm
Max 1.00 m

Going

Landing

Rise

Maximum pitch 42°
Minimum going 220 mm
Maximum rise 220 mm
Baluster gap should not let a 100 mm diameter sphere pass through
Twice rise plus going should fall between 550 mm and 700 mm

Figure 3.65 Requirements for domestic staircases

Figure 3.65 details the requirements for domestic staircases. Staircases in other buildings may have different requirements particularly with respect to fire resistance, pitch angle and tread and riser sizes.

Stairs may be arranged in different layouts (Fig. 3.66), common ones are:

- straight flight
- quarter turn
- half turn
- dog-leg
- open well
- quarter turn with winders

Constructional details of a timber staircase are shown in Fig. 3.67. In some cases designers prefer to omit the risers which creates a space and the term **open riser staircase** is used. A simple concrete staircase is shown in Fig. 3.68. Concrete staircases are used when:

- high loadings have to be carried
- fire resistance is a requirement
- stair widths are greater than can be economically achieved in timber
- buildings are being constructed from in-situ concrete
- staircases are in public buildings

Stability of timber stairs is maintained by:

- the strings supporting the tread and risers
- newel posts supporting the strings to landings
- glue blocks strengthening tread and risers

Stability of concrete stairs is maintained by:

- using the correct strength of concrete
- properly placed reinforcement
- maintaining minimum waist dimensions
- designed connections to landings

Fire resistance of timber stairs is minimal. Slight protection is sometimes provided by the underside of the stair being under drawn with plasterboard and skim. Greater protection to timber stairs is provided when they are within a protected fire escape route in properties of multiple occupation.

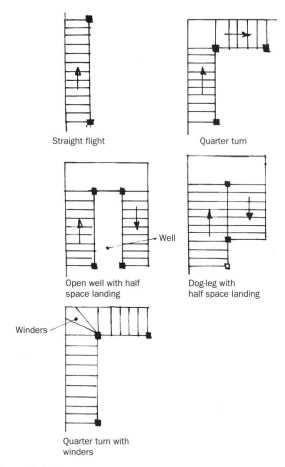

Figure 3.66 Stair layouts

Hardwoods are, however, dense timbers and resist abrasion well. Non-slip pads are sometimes applied to the tread surface.

Concrete stairs are extremely durable whether the treads are left uncovered or have a decorative facing applied to them. Such facings might be:

- PVC tiles
- terrazzo tiles
- carpet
- hardwood tread

Access and egress The pitch angle of the stairs dictates that they should not be too steep or too shallow in their pitch, likewise the relationship of the tread and riser and the nosings falling into line ensures that tripping on the stair does not arise. The handrail at the correct height coupled with the balustrading ensures support and safety for those using the stairs.

Headroom between the stairs and the bulkhead ensures that all people can use the stairs without fear of bumping their head. Likewise large items of furniture can be transported up or down the stairs with relative ease.

Landings at the top and bottom of every flight and sometimes when used as intermediate landings ensure that there is room to join and leave the staircase in a safe manner. They also provide a circulation space which allows people or objects to pass safely.

Landings also allow for a change in direction of the stair to be achieved safely. No door is allowed to open onto a landing position unless it is a cupboard or access duct door or at the bottom landing. In all of these cases a clear space of 400 mm must remain across the width of the landing.

Concrete stairs provide fire protection which will be more than sufficient to allow building users to escape before the stairs collapse. It is important that the 25 mm depth of cover to the reinforcement is maintained when casting the stairs to maintain the strength of the stairs.

Durability of timber stairs is provided for by using parana pine for the treads and strings. In most cases softwood stairs are covered with carpet which affords further protection against wear. Hardwood is also used for stairs and may be left uncovered.

Self-assessment tasks

1. Make a sketch to show the following stair terms: tread, riser, pitch, pitch line, headroom, nosing, newel, glue block, string.
2. Explain the difference between a quarter turn landing and a quarter turn with winders.
3. Why is the dimension of the 'waist' important on a concrete stair?
4. Why do landings have minimum sizes?
5. State the critical dimensions for guarding a staircase and a landing.
6. Prepare a detail of a straight flight of stairs and illustrate on it the requirements of the Building Regulations Approved Documents.

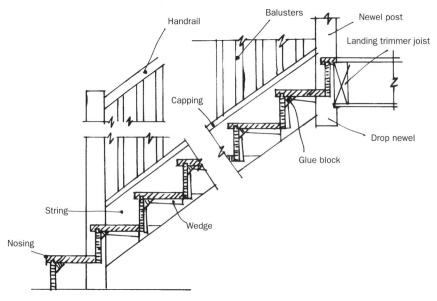

Figure 3.67 Timber staircase

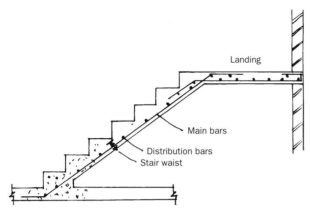

Figure 3.68 Concrete staircase

Function of finishes

Building exteriors and interiors are often judged by their aesthetic appeal in a somewhat subjective manner by the observer. While there is no doubt that the aesthetic appeal of a finish might contribute to its choice, there are other factors which need consideration. These are:

- durability
- impact resistance
- hygiene factors
- resistance to moisture and condensation
- thermal stability
- light reflection or absorption
- freedom from slipperiness
- thermal insulation
- fire resistance

These factors will vary in importance according to the situation in which they will be used, for example:

- kitchens
- bathrooms
- lounge
- office
- warehouse
- workshop
- changing rooms
- display areas
- reception areas
- storerooms

What is common for internal locations is that they all require a choice of wall, floor and ceiling finish. External to the building, wall and roof finishes in particular are required, which may be coupled with a 'finish' for access and parking areas.

Various classifications can be used for finishes but those requiring consideration here are:

- wet finishes
- dry finishes
- self finishes

Wet finishes

The most common finishes under this heading are:

- screeds used on floors
- renders used on walls
- plaster used on walls and ceilings

Screeds

Screeds are an applied finish to floors. Their purpose is to:

- provide a smooth surface to levels or gradients
- to enable services to be accommodated, e.g. electrical conduit
- provide a sub-base for other finishes, e.g. floor tiles
- act as floor finish in their own right, e.g. granolithic screed
- in some cases act as part of the load-bearing floor, e.g. structural screed

Screeds are normally made from either a sand and cement mixture or a fine concrete. The thickness of the screed is related to the type of base on which it is to be laid. Factors which can affect this are:

- structural strength of the floor
- the degree of bond to the sub-base

Four classes of screed are:

- monolithic
- bonded
- unbonded
- floating

Monolithic screeds are normally trouble-free, being laid directly on to the in-situ concrete floor within three hours of it being placed. This minimises the degree to which separation, cracking and curling occurs through differential shrinkage. As this screed is laid early in the construction process its surface will require temporary protection throughout the contract.

Bonded screeds are laid after an in-situ floor has been completed and cured. The slab surface will need to be hacked or scabbled to expose the aggregate. Then a cementitious grout or chemical adhesive is used to assist the applied screed to bond to the surface of the slab. A minimum thickness of 25 mm is used but 40 mm is preferred.

Unbonded screeds need to be at least 50 mm thick to maintain the stability of the screed and reduce movement or the possibility of loads breaking the screed. Underfloor heating will require an increased screed thickness.

Floating screeds are unbonded screeds laid over layers of fibre board or glass fibre quilt. This enables floors to possess improved sound and thermal insulation properties. To maintain stability and strength thicknesses of 65 mm to 100 mm are used.

Renders

Render is a term used to describe a cement based coating which can be applied to internal or external walls. Renders are able to:

- create or accept a variety of surface finishes
- increase resistance to moisture
- improve thermal resistance of walls
- resist quite high impact loads

Renders once cured will move very small amounts, however external renders will be subject to a greater variety of temperature and moisture changes. Materials used for renders are:

- Portland cement
- lime

- sand
- aggregates, mainly for aesthetic qualities

Sand will influence the performance of the finished render. Coarse sands are used for the undercoat and graded sands for the final coat. The choice of mix is influenced by:

- background materials to which the render is being applied
- local environmental conditions

Table 3.12 Typical mixes for renders

Render type	Mix proportions
Cement/sand with plasticiser	1 : 3
	1 : 5
	1 : 7
Masonry cement/sand	1 : 2.5
	1 : 4
	1 : 6 : 5

Plaster

Plaster is an applied finish to walls and ceilings. It is also able to hide variations in surface finishes in the background to which it is applied. Normally on walls two coats are applied; an undercoat and a final coat. Characteristics of plaster surfaces are:

- smooth surface
- jointless
- hygienic
- crack-free

In addition plaster surfaces can accept a variety of applied finishes such as:

- wallpaper
- emulsion paint
- gloss paint
- ceramic tiles
- acoustic tiles
- spatter dash finishes

Plaster finishes are also used to enhance the following properties of a background:

- thermal resistance
- acoustic qualities
- fire resistance

The choice of a particular plaster finish is dictated by the type of background, for example:

- brick
- thermal block
- concrete
- plasterboard
- stone
- expanded metal lathing/mesh

These backgrounds will then possess certain qualities which affect the choice of plaster. These are:

- strength
- suction
- bonding characteristics
- degree of movement
- presence of water soluble salts

The most common plaster classification is Class B retarded hemi-hydrate gypsum plaster, and can be used for general plastering applications. Final coat plasters are finer than undercoat plasters. Four classifications occur within Class B:

Undercoat plasters:

- Browning
- Metal lathing

Final coat plasters:

- Finish
- Board finish

Lightweight plasters are a mixture of Class B gypsum plaster and either exfoliated vermiculite or expanded perlite. **Bonding plasters** that have been mixed with vermiculite are suitable for ensuring a good bond to concrete surfaces. **Board finish plasters** are used on plasterboard as a single finish coat.

Dry finishes

These are finishes which are usually applied to a background. Examples of these finishes are:

- plasterboard
- timber panelling
- laminate sheet
- tiles

Plasterboard

Plasterboard is made from a solid core of gypsum bonded to two face sheets of durable lining papers. The board is fixed to walls and ceilings as a dry lining, or as a base for gypsum plaster finish. Fixing is normally by galvanised nails to timber or plaster dab and dots. The boards are not suited to damp humid conditions.

Wallboard is a plasterboard **dry lining** for internal applications. Its use allows:

- a reduction in the amount of wet trades work and drying out time
- spaces to be more responsive to space heating
- surface condensation to be reduced
- vapour transmission to be controlled

Additional advantages are improving sound, thermal and fire resistance. One face will allow direct decoration to the surface and the other will accept a coat of finish plaster.

Timber panelling

Aesthetic variations to wall surfaces can be created by using either solid timber panels contained within a framework or plywood panelling. The latter would have a surface veneer of an exotic or rare hardwood. Plywood panelling is preferred as it is a stable material and can be used in internal and external locations.

The panelling is either fixed to a sub-frame or to a framework of timber grounds. Whichever is used it must be remembered that services and loads from surface fixings will need to be accommodated. In order to prevent the possibility of forming ideal conditions for dry rot, the space behind the panelling and the wall should be ventilated.

Laminate sheet

An alternative to hardwood veneers is the use of plastic laminate sheet applied to a board background. Laminate sheets are available in a range of colours and can have printed patterns and other artwork incorporated in the sheet. The laminate surface can provide very good resistance to:

- dry heat up to 180 °C
- staining
- moisture
- cigarette burns
- steam
- surface crazing

The sheets are also easy to keep clean and are non-tainting to foodstuffs.

Tiles

Tiles are an applied finish to walls and floors of buildings in either internal or external locations. Typical applications are:

- entrances to buildings
- en-suite and bathrooms
- shop floors
- restaurants
- corridors

Interior glazed wall tiles are available in square, rectangular and hexagon shapes with either square or rounded edges. At junctions internal and external angle bead tiles are available to provide neatly radiused corners. The tiles are fixed to wall surfaces by a thick or thin bed of proprietary adhesive. In some situations any background movement may cause the tile surface to fail. In areas where the tiles are exposed to water a solid bedding of waterproof adhesive should be used together with waterproof grout or epoxide resin. These tiles are:

- smooth or textured on the surface
- resistant to water
- resistant to chemical cleaners
- available in a range of attractive colours
- dimensionally stable under most conditions
- easily cut to awkward shapes
- resistant to surface abrasion

Floor finishes include the following types of tiles:

- quarries
- cork
- plastic

Quarry tiles are produced in a variety of shapes and sizes together with several 'special' shapes. Colours available include reds, blues, browns and buff. Surfaces which reduce slipping are available. They are bedded on to a firm sub-base with either proprietary or cement-based adhesives. Joints between the tiles are grouted.

Typical uses are:

- kitchens
- entrance halls
- dairies
- corridors
- bakeries

Cork floor finishes are available in tile and sheet form and are suitable for all applications except where very heavy wear is anticipated. Cork can be laid on any level floor with a damp proof membrane. Cork possesses the following characteristics:

- good wearing qualities
- sound absorbent
- non-slip
- warm to the feet

Maintenance of the cork surface is limited once the surface has been sealed. Different densities of tile are available but they will not resist surface indentation from heavy objects placed on, or dragged across the surface.

Plastic tiles are as thin as 1.5 mm and require the following characteristics from the sub-floor:

- stable and rigid surface
- smooth with no projections
- free from damp or moisture

Polyvinyl chloride (PVC) is used in the production of all types of tiles and sheet and has proven performance and durability. The tiles can be laid on most bases with a bituminous adhesive. Tiles can join to each other accurately and are easily cut around awkward shapes and objects such as pipes. These tiles are not suited to industrial kitchens.

Self finishes

There are some materials which are used in the construction of walls and floors which can produce a surface finish requiring no further modification by the application of another material. These are known as **self finishes** and include:

- brickwork
- concrete

Bricks

Brickwork which has no applied finish is known as **fair face brickwork**. Fair face walls may be constructed in the following bricks:

- commons
- facings
- engineering

A full range of brick colours is available with the main ones being red, yellow, blue, brown and multi-colour. Various types of mortar colours may also be used to enhance the brickwork. Designers will often specify certain types of brickwork bond which will add to the aesthetic quality of the wall. Bricks may also be indented or projecting which will produce interesting shadow patterns in sunlight.

The surface texture of bricks is also used to vary the appearance of a wall. Typical surface textures available are:

- smooth
- sand faced
- horizontally dragged
- stippled
- rusticated
- sand creased

Situations where fair faced brickwork can be found are:

- internal and external walls
- street furniture

Figure 3.69 In-situ profiled concrete panel

- garden planters
- paths and staircases
- fire protection to structural steel frames
- sculptures

In-situ concrete

The fluid nature of concrete before curing can allow it to be formed into a range of shapes and sizes both on and off the site. Profiled shapes called formwork are used to support the finish until the concrete has cured. These are direct finishes and will require strict control on the formwork production, concrete casting and removal of the formwork. Typical direct finishes are:

- smooth – from planed boards, plywood or steel sheet
- rough – emphasising the grain from sawn boards
- raised joints – using chamfered edge boards
- recessed joints – using various cross-sections of timber fillets (Fig. 3.69)

Formwork may have linings of rubber or thermoplastic sheet applied to them which can produce a full range of textured, profiled and patterned surfaces. In some situations the designer will specify an exposed aggregate finish to the concrete surface. This can be achieved by using a retarder in the casting process which then allows the cement grout to be removed, or by allowing the concrete to cure and then exposing the aggregate with a scabbling tool.

Self-assessment tasks

1. Briefly describe the essential differences between the three types of screed.
2. State how the performance requirements for a floor and wall tile might differ.
3. Explain how brickwork and concrete can be used to provide decorative finishes.
4. What are the advantages of dry lining a wall?
5. What functions does render perform when applied to an external wall?
6. Create a matrix to illustrate the different wall, floor and ceiling finishes used in your college.

3.3 Primary services in low-rise domestic buildings

Primary services for low-rise domestic buildings

It can be seen from studies earlier in this unit that the type of activity within the building can influence not only its overall size and shape but also how it might be divided up inside into smaller areas which are likely to have designated uses. The type of activity and the use of the space can then in turn influence the building services which may be installed, and their distribution.

Building services are required because the occupants and other users of the building will expect one or more of the following to be provided in order to satisfy their needs.

It is not uncommon therefore to find the following services:

- water
- gas
- electricity
- drainage
- space heating
- telecommunications

Self-assessment tasks

Before reading on write brief notes describing the purpose to which the following are used within buildings.

1. Water
2. Gas
3. Electricity
4. Drainage
5. Space heating
6. Telecommunications

Functions of primary services

Water supply

Water is an important primary service as it not only provides a ready supply of cold water into a property for drinking purposes but also it can be:

- warmed to provide hot water supplies
- heated to provide space heating
- used as a means of removing various types of waste
- incorporated into industrial processes
- used as a liquid feed for plants

British Standard 6700 relates to the design, installation, testing and maintenance of services supplying water for domestic buildings, and British Standard 5546 to non-storage appliances. Water supply is controlled by the use of Model Water Byelaws. As all water supplied from the water main is treated to a standard fit for drinking by humans the byelaws are aimed at preventing water from being:

- wasted by flowing away unused from an appliance or through leakage
- unduly consumed by using more water than is required for a different purpose
- misused for purposes other than for which it was supplied
- contaminated through use by any means

Gas supply

Gas is normally supplied directly to properties from the gas main to buildings though an alternative source is that of liquefied petroleum gas (LPG). Gas supplies enable the provision of one or more of the following:

- hot water
- space heating
- cooking facilities
- processing requirements (for example heat treatment of metals)

Gas installations are controlled by the Gas Safety (Installation and Use) Regulations 1984 which amended the Gas Safety Regulations 1972 and the Gas Safety (Rights of Entry) Regulations 1983.

Electricity supply

Electricity is generated at power stations and fed into the national grid system and from there to substations for distribution to cities, towns, industrial estates, large individual users and groups of villages. There is no fixed pattern for local distribution, this being developed as a result of the requirements of the area.

Electricity supplies are used to provide power for:

- lighting circuits
- water heating
- space heating
- operation of machinery
- small electrical appliances

Electrical installations are controlled principally by the Electricity Supply Regulations 1988 (Amended 1990) and the IEE Regulations for Electrical Installations 16th Edition.

Drainage installations

The provision of water supplies into a property creates the need to eventually dispose of it efficiently and effectively. This is irrespective of whether or not the water is contaminated. Likewise rainwater falling on the building or the surfaces surrounding it will need to be collected and discharged in a safe manner. Drainage installations therefore deal with the collection and discharge of:

- surface water
- foul water

Drainage installations are controlled by the Building Regulations Part H and the relevant Approved Document. Other principal publications which relate to and influence drainage systems are British Standard 8301: 1985 and Building Research Establishment Digests.

Space heating

Buildings must provide satisfactory levels of thermal comfort not only for their occupants but also for the mechanical systems they accommodate. Thus buildings will be heated to satisfy the demands of:

- health and comfort of people
- health and comfort of animals
- production processes in manufacturing
- storage of goods
- horticulture

Space heating installations will need to be designed to provide agreed levels of thermal comfort. The principal guide to the performance required can be found in the Chartered Institute of Building Service Engineers (CIBSE) Design Guide. There are many different methods by which spaces can be heated and therefore installations can be subject to the legislation and standards which relate to water, gas, electricity and drainage, some of which have been identified previously.

Self-assessment tasks

Identify the relevant standards associated with:

1. Water supply
2. Gas supply
3. Electricity supply
4. Drainage installations
5. Space heating

Cold water supplies

Water supplies taken from the mains supply to a property are permitted provided that the installation complies with the Water Act 1989, and in addition that the whole installation complies with the current water byelaw requirements. British Standard 6700: 1987 specifies requirements for design, installation, testing and maintenance of services supplying water for domestic use within buildings.

Connection of water supply to a property

Buildings in urban areas will take their water supply directly from the water main (Fig. 3.70). It is normal practice to make a connection to the main without isolating the supply. A mains tapping plug cock is used which allows the communication pipe to the property to be connected. The communication pipe is that part of the service pipe which is owned by the water authority. The service pipe delivers water to the building at mains pressure, and is controlled by a stop valve. This valve can be used to control the flow of water into the property.

Concern for the conservation of water, and an increase in

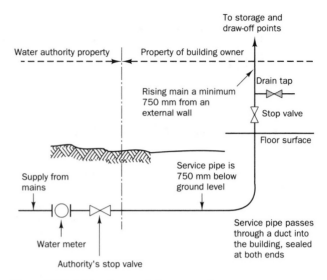

Figure 3.70 Cold water supply to a property

consumer demand has led to the water authorities considering and in some areas installing water meters which allow for the accurate measurement of water consumption. They may also help with the tracing of leaks. The meters are usually situated outside the boundary of domestic premises, but may be within industrial and commercial buildings.

The position and depth of the service pipe should be such that it is not affected by:

- frost
- heavy vehicular traffic
- the building load

Where the service pipe passes through into the property it should be housed in a protective duct. This will aid replacement of the service pipe when necessary. Suitable materials for service pipes are polythene and PVC. As the service pipe enters the property a stop valve and drain down provision are made for the consumer. This allows the consumer to isolate the water supply from within the premises in an emergency, and drain down the internal pipework when required. In addition the water byelaws require that a non-return valve is fitted at the intake position in order to prevent possible back flow of contaminated water into the water main.

Installation, distribution and operation of cold water systems

Cold water distribution within any building must always satisfy the requirements of the water byelaws. It is important that the quality of the drinking water is maintained up to its draw-off point and that the mains supply is also protected from becoming contaminated.

Correct installation and distribution should ensure that the water supply is efficient in operation. Systems should be installed in such a way that they will:

- prevent contamination of the water main
- not become contaminated by materials used in the system
- allow for isolation and draining down
- not be affected by movement of the building
- not be unduly affected by low temperatures
- prevent stagnation of the water
- allow access for maintenance

Distribution within the building should ensure that:

- the cold water is delivered at an acceptable temperature
- design flow rates are provided and maintained
- drinking water is provided directly from the main
- noise levels are kept to a minimum
- economical pipework layouts are used
- adequate support is given to pipework and storage cisterns
- appliances are connected in accordance with byelaw requirements, e.g. air gaps
- isolation of appliances or delivery zones is possible

The indirect system

This is the preferred method as it provides fewer opportunities for the mains supply to become contaminated. This is because the number of draw-off points taken directly from the incoming rising main will be limited (Fig. 3.71). Normally these draw-off points will only serve outlets which provide drinking water, or are used in the preparation of food.

The following are characteristics of the indirect system:

- drinking water only is fed from the rising main
- supplies to all other outlets are taken from a cold water storage cistern
- a reserve of cold water is available in cases of failure or isolation of the mains supply
- less noise due to water flow
- most of the fittings operate at a pressure less than the main
- peak demand on the main is reduced
- distributing pipework may be larger in diameter
- risk of mains contamination is reduced

The direct system (Fig. 3.72)

The open-vented system is not commonly installed today, however the student engaged in maintenance, renovation or alteration of buildings may come across it. The system provided all outlets requiring cold water with a feed directly from the rising main. This could have increased the risk of contamination back to the water main.

Now gaining in popularity is the unvented mains fed system (Fig. 3.73). This is usually installed together with a mains fed unvented hot water supply. Precautions are necessary to prevent contamination of the water main and to reduce the pressure of the main within the distribution pipework and fittings.

Characteristics of the system are:

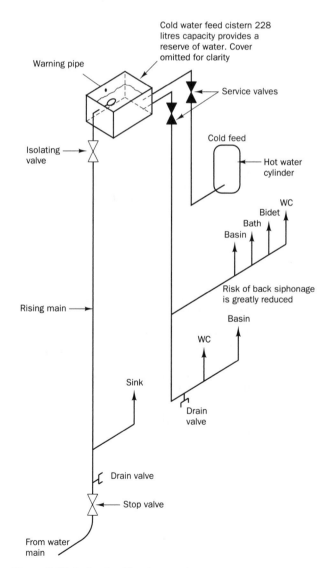

Figure 3.71 Indirect cold water supply

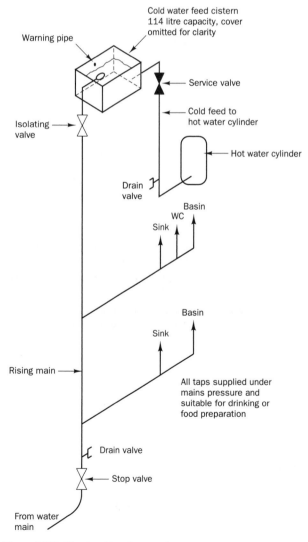

Figure 3.72 Direct cold water supply

Figure 3.73 Unvented mains fed cold water supply

- all fittings and appliances are served from the main
- all cold water taps will supply drinking water
- there is no cold water storage when mains supplies fail
- fittings will be subject to constant pressure
- noise due to water flow is possible
- less pipework may be used than in the indirect system

Components and fittings for cold water systems

The operation of a cold water installation cannot function without the use of fittings and components. These are shown in Table 3.13.

Table 3.13 Fittings and components used in water supply systems

Component/fitting	Purpose
Stop or globe valve	To regulate water flow and provide a means of isolation, normally fitted to the rising main
Non-return valve	To prevent the back flow of water into the main
Gate valve	To control or isolate water flow on the indirect cold water supply; acts as a service valve
Drain valve	Allows for the whole or part of a system to be drained down
Ball valve	Provides an automatic means of controlling the supply of water to a cistern
Taps	Used to control the flow of water at the outlet point
Pipework	Used to convey the water, can be copper, mild steel or PVC
Cistern	Used to store cold water prior to use, can be steel, GRP or polythene
Pressure reducing valve	Reduces mains pressure to acceptable design pressures for the system

Self-assessment tasks

1. Describe the differences between an indirect cold water installation and an unvented mains fed system.
2. How can freezing of the cold water service pipe be prevented?
3. What is used to prevent contamination of the water main from a property?
4. Explain the purpose of the following: cistern, gate valve, stop valve, ball valve.
5. How can wastage of cold water be reduced?
6. Prepare a collage which illustrates the fittings and components used in water supply systems together with their purpose.

Hot water supplies

It is important that hot water is supplied in sufficient quantity and at an adequate temperature in order to meet the needs of the user. In domestic properties water at the kitchen sink may be up to 60 °C for washing up purposes, whereas at other outlets the temperature can be at 40 °C. Flow rates will vary from 0.1 litres/sec to 0.3 litres/sec for the shower or bath respectively. Hot water supplies in domestic premises are likely to be supplied in two different ways. The water supplied will be either from:

- a **local system**, or
- a **central system**.

Local systems

Local systems are often referred to as instantaneous systems, where cold water is heated as it passes through the appliance on its way to the draw-off point, the water flowing and being heated for as long as the outlet tap is open. Instantaneous systems do not have the ability to store reserve supplies of hot water.

The amount of hot water available is unlimited, but the flow rate is usually slow as the water temperature is dependent upon flow rate and heat energy input. These heaters can therefore only provide hot water for draw-off and not supply water for central-heating systems (Fig. 3.74).

Gas or electricity may be used as energy sources to either single or multi-point instantaneous heaters, single point normally serving one appliance such as a sink and multi-point serving several such as a bath, basin and sink. Instantaneous heaters may also be used to provide hot water for shower outlets.

Central systems

Central or storage systems involve the cold water being heated and stored some time before it is needed. Once the water has been used the cold water which replaces it has to be reheated to

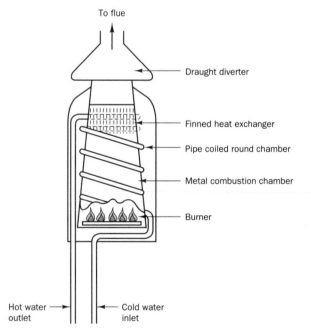

Figure 3.74 Instantaneous gas water heater

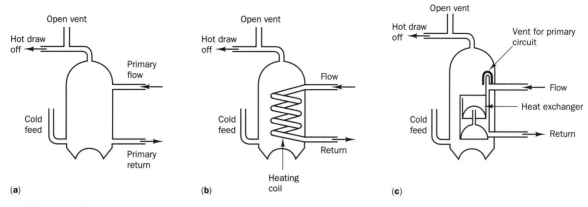

Figure 3.75 Types of hot water cylinder: (a) direct; (b) indirect; (c) primatic

the required temperature before being drawn off again. The time taken for this to occur is called the recovery period.

The quantity of water stored and the temperature that it is heated to will be designed to be sufficient to meet the needs of the appliances relative to the type of building use. Central systems have the advantage that water can be drawn off at a constant temperature and flow rate.

The principal components (Table 3.14) of central systems are the:

● boiler which raises the temperature of the water
● cylinder which stores the heated water
● cistern which stores cold water in order to replace the hot water which has been drawn off

Central systems may be:

● vented
● unvented

Vented systems are fed from a cold water storage cistern. This is necessary to provide sufficient head of pressure in the system and to accommodate any expansion of water when it is heated in the boiler. Unvented systems do not use a storage cistern, but rely on a direct feed from the incoming cold water supply. Special precautions are needed with this type of installation to prevent contamination of the water mains.

Vented systems

These are often classified as:

● direct
● indirect

These terms relate to the type of hot water storage cylinder which is used. The two principal types are called direct and indirect and have a capacity of 120 litres. Direct cylinders allow water from the boiler to mix directly with the water which is being heated, whereas indirect cylinders separate the water from the boiler or primary circuit and the hot supply to the outlet points on the appliances (Fig. 3.75). Indirect cylinders possess a heat exchanger which prevents the build-up of scale deposits, and maintains the efficiency of the cylinder.

It is important that the cylinders are insulated with factory applied insulation that restricts standing heat losses to 1 watt/litre. Normally a coating of PU foam 35 mm thick and density 30 kg/m^3 will satisfy this requirement. A thermostat should be fitted to the cylinder to control the temperature, together with a timer to control the heating times.

Direct systems The water heated in the boiler circulates to the direct cylinder where it mixes directly with, and raises the temperature of the stored water. Hot water rises to the top of the cylinder where it is then drawn off. Cold water then replaces that which has been drawn off. Circulation of water in the primary circuit may be by natural convection currents or speeded up by using a circulator or pump (Fig. 3.76).

Indirect systems Heated water from the boiler is separated from the hot water draw-off by the heat exchanger in the indirect cylinder (Fig. 3.77). It will be noticed that a separate cold feed and expansion tank is required for the boiler and primary circuit. This system is ideal for supplying heated water for small bore central-heating systems.

Unvented systems

These systems are fed directly from the main and remove the need for a cold water storage cistern (Fig. 3.78). Cylinders

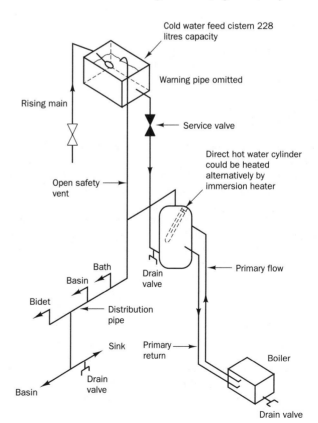

Figure 3.76 Direct hot water system (vented)

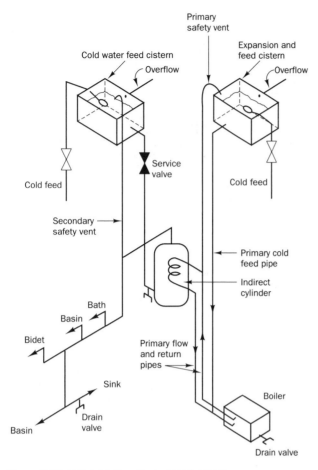

Figure 3.77 Indirect hot water system (vented)

may be direct or indirect. Primary and secondary circuits are now pressurised which means that safety precautions have to be installed to:

● prevent explosion
● accommodate expansion of the heated water
● control the water pressure

Immersion heaters

Immersion heaters use electricity to heat an element inserted into the hot water cylinder (Fig. 3.79). The immersion can be used as an independent heat source or to provide supplementary heat to storage systems. Immersion heater elements may be single or dual.

Dual immersion heaters consist of two independent thermostatically controlled different length elements, mounted on a single head. The longest element is selected for water for a bath and the shortest for a basin, or quick boost. The elements are usually a 3 kW rating.

Self-assessment tasks

1. Briefly describe the features of an instantaneous and central system of hot water supply.
2. Why is an indirect cylinder used?
3. What would be the effect of a low flow rate at the bath tap?
4. Explain the purpose of the following: cylinder, boiler, thermostat, circulating pump.
5. Why are dual immersion heaters used?

Gas supply

Gas supplies are brought to domestic buildings through a service pipe which connects the gas main in the street to the consumer. Service pipes normally enter buildings on the elevation closest to the main and without passing under the foundations of the building (Fig. 3.80).

It is common practice to install meter boxes in the outside walls of buildings, but where the service pipe must pass through the structure it must be sleeved and sealed. Pipes on the face of walls will have to be securely fixed.

Once inside the property the meter installation will include a gas cock, governor, filter and a meter. It is also practice to have a pressure test point (Fig. 3.81).

Table 3.14 Fittings and components used in hot water supply systems

Component/fitting	Purpose
Cylinder	Stores the heated water before use
Immersion heater	A form of heating element immersed in the cylinder and used to raise water temperature
Boiler	A source of heating water
Heat exchanger	Used in cylinders to provide efficient heat exchange and to separate primary and secondary circuits
Thermostat	Controls water temperature either in the boiler or in the hot water cylinder
Timer	Used to control the operation of the boiler or the immersion heater
Thermal insulation	Reduces the rate at which heat is lost from pipework or the cylinder
Circulating pump	Used to increase flow around systems and overcome gravity
Instantaneous heater	A form of water heater that raises the temperature of water as it flows through it
Vent and expansion pipe	Prevents a system from becoming pressurised and allows water to expand and contract on heating and cooling
Expansion vessel	Accommodates an increase in volume of heated water to reduce burst pipework and danger of scalding from hot water.

Table 3.15 Fittings and components used in domestic gas supply

Component/fitting	Purpose
Service pipe	The pipe from the main to a property
Gas cock	Allows the gas supply in the property to be isolated from the main, or an appliance in the property from its supply
Governor	Reduces the pressure from the main to a pressure more suited to the appliances being served
Filter	This collects the fine particles which could clog the gas jets of appliances
Meter	Records gas consumption
Pressure test point	Allows a check to be made on pressures and the governor efficiency
Meter box	Normally fitted in the external wall and contains the meter, governor and control cock

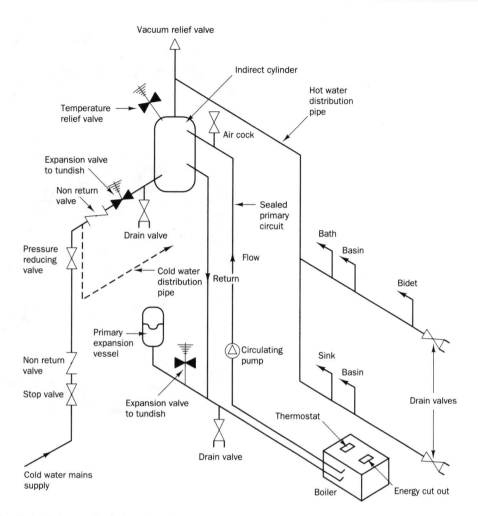

Figure 3.78 Indirectly heated unvented hot water system

Electricity supply

Electricity supplies enter a building at the intake position. A service cable brings the supply to a meter box installed in the external wall. The cable will be ducted into the box (Fig. 3.82). The cable terminates at the sealing chamber which contains the service fuse. A connection is then made to the meter from which cable tails are connected to the consumer unit and isolation switch (Fig. 3.83).

Distribution of electricity in the property

Distribution of electricity in domestic properties is normally in the form of circuits which serve power and lighting requirements. Table 3.16 shows typical fuse ratings and applications.

Power circuits

Two types of circuit are permitted:

- **ring circuit**
- **radial circuit**

Ring circuits use twin and earth wire looped from one socket to another with both ends being connected at the consumer

Table 3.16 Typical fuse ratings and applications

Fuse ratings (amps)	Application
32 or 50	Cooker
30	Ring circuit ground floor
30	Ring circuit upper floor
30 or 32	Radial circuit kitchen
15	Immersion heater
15	Garage
5	Central heating controls
5	Lighting ground floor
5	Lighting upper floor

unit to the same 30 ampere fuse or circuit breaker (Fig. 3.84). This allows any number of sockets and fused connection units per 100 square metres of floor area. In a house it would be split into two circuits, one serving upstairs and the other downstairs. Spurs may be added provided that their number does not exceed the total number of socket outlets, and that there is no more than one single outlet, one twin socket outlet or one fixed appliance per spur.

Socket outlets in bathrooms are forbidden by the IEE Wiring Regulations. Exceptions to this are shaver sockets and safety extra low voltage sockets. Other fixed electrical equipment in the bathroom must not have switches within reach of the person using the bath or shower. This is

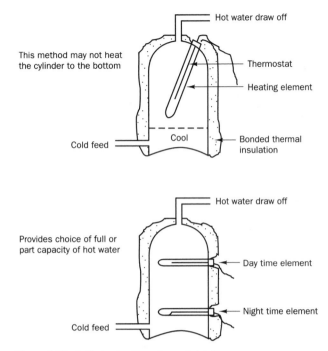

Figure 3.79 Cylinder heating element detail

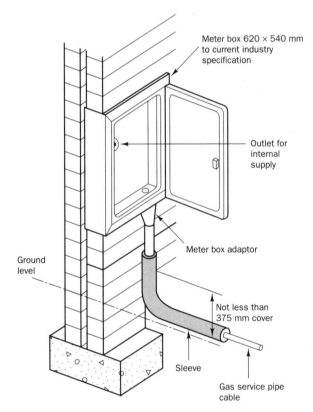

Figure 3.80 Gas supply to a property

Figure 3.81 Gas meter box

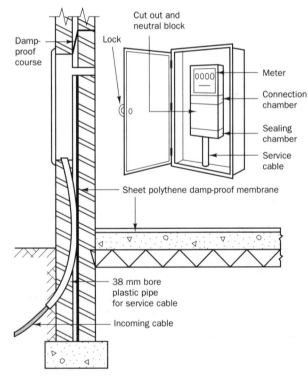

Figure 3.82 Electrical supplies to a property

overcome by using pull cord switches. Correct earth bonding is also required for exposed metal parts in bathrooms.

Radial circuits may serve an unlimited number of BS 1363 socket outlets provided that the area does not exceed 50 square metres and they are wired in 4 mm² PVC cable with a 30 ampere or 32 ampere cartridge fuse or circuit breaker respectively. For areas up to 20 m² the circuit protection is 20 amperes with 2.5 mm² PVC cables. Any type of overcurrent protective device may be used (Fig. 3.85). Radial circuits are commonly used for showers, immersion heaters, cookers, and instantaneous water heaters.

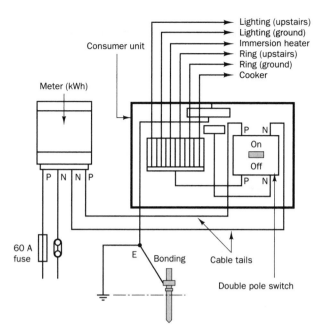

Figure 3.83 Domestic distribution board

to earth. This bonding will ensure that all metalwork in the building will be at the same potential (i.e. no voltage difference) in the event of metalwork becoming live to earth.

Table 3.17 Fittings and components used in electricity supply

Component/fitting	Purpose
Service cable	The cable connecting the property to the main
Service fuse	Protects electrical supplies in the street
Meter	Records electricity consumption
Cable tails	Cables used to connect the meter to the consumer unit
Consumer unit	A box which incorporates the mains isolation switch and individual protective devices for circuits in the property
Isolation switch	A double pole switch which controls the phase (live)

Self-assessment tasks

1. Explain the purpose of the gas cock, governor and meter.
2. What precautions must be taken when the gas supply pipe passes through an external wall?
3. Describe the difference between a ring and a radial circuit.
4. Select four different electrical appliances in your house and state their correct fuse ratings.

Lighting circuits

For domestic properties lighting circuits are wired either using the 'loop in' method or by using a joint box incorporating four terminals or a block connector strip (Fig. 3.86). These will be served from the consumer unit to individual radial lighting circuits. Various switching arrangements are possible to suit the convenience of the user, a common one being two way switching to a staircase, large room or corridor. Again two circuits are normally installed in a house so should one circuit fail some part of the property would have some light.

Earthing

Any earth fault current will try to find a low resistance path to earth. All services must therefore be bonded as near as possible to their entry into the building. Likewise the exposed metal surfaces of radiators, baths, sinks and tanks must be bonded together and

Drainage installations

The use of water with various fittings and appliances in buildings creates a need for the efficient collection and discharge of the foul and waste water whether it is contaminated or not. Similarly rainwater which falls on the building or the surfaces surrounding it, or the water in the sub-soil will need to be collected and discharged in a safe manner. For drainage purposes water may be classed as:

- **surface water**
- **foul water**

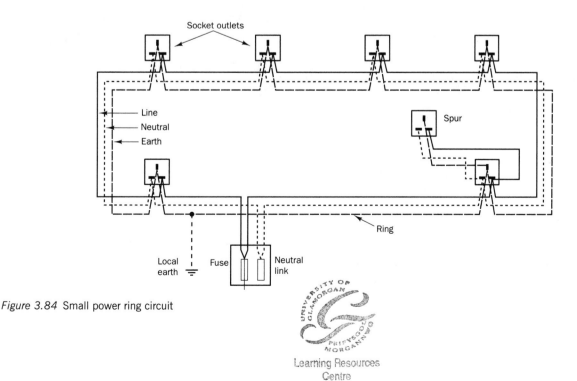

Figure 3.84 Small power ring circuit

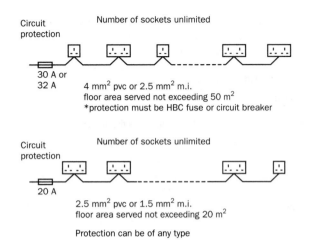

Figure 3.85 Radial power circuit

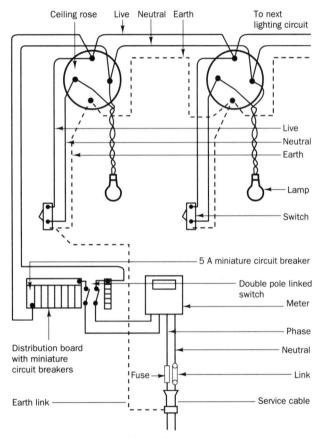

Note: A ceiling rose shall not be installed in any circuit operating at a voltage normally exceeding 250 volts (IEE Regs)

Figure 3.86 Loop in method for lighting circuits

Surface water

Surface water may be collected from the roofs and faces of buildings, and the surface areas around them such as drives and paved areas. This water is normally considered to be clean and therefore its discharge may be directly to main sewers, convenient watercourses or soakaways.

Soakaways collect and store rainwater during a storm and then release it into the surrounding ground as quickly as the ground conditions will allow. The rate of this discharge is also influenced by the height of the water table.

Soakaways are normally sited a minimum of 5 metres away

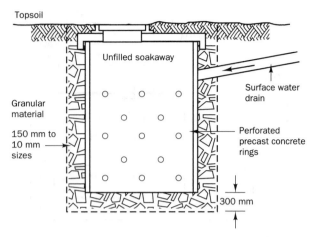

Figure 3.87 Concrete soakaway

from a building to reduce possible settlement of the building's foundations. Small soakaways are filled with coarse granular material and hardcore. Larger soakaways are formed from dry jointed perforated concrete rings (Fig. 3.87). The outside of the ring is backfilled with granular material to allow the water to percolate into the surrounding ground more effectively. The soakaway is covered with a concrete slab which will also allow access should it be needed.

Sub-soil drainage

Sub-soil drainage is provided to prevent the passage of moisture beneath a building, and to reduce possible damage to the fabric of the building. Sub-soil drainage will also increase stability of the ground and its horticultural properties. A variety of methods are available, these being illustrated in Fig. 3.88.

Foul water

Foul water is waste water from a sanitary convenience or other soil appliance, or water which has been used for washing or cooking. Building Regulations require that any

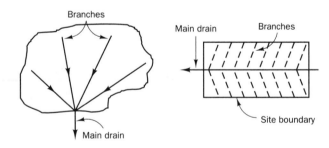

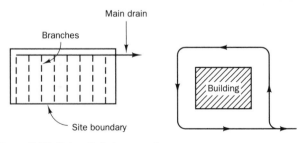

Figure 3.88 Sub-soil drainage systems

system which carries foul water should be adequate. This applies to drainage systems both above and below ground. In this respect a drainage system should:

- carry the foul water to a suitable discharge point
- minimise the risk of leaks or blockages
- prevent foul air from entering the building under normal operation
- be ventilated
- provide access for clearing blockages

Above ground drainage

This is commonly called sanitary pipework and should be designed, installed and maintained so that it minimises the risk to health that could arise through leakage, blockage or surcharge of the system as a result of discharge from waste and soil appliances.

In domestic properties soil and waste appliances are discharged to a single stack. The single stack (Fig. 3.89) is used in properties where the appliances are grouped closely together, and can be connected to the discharge stack to assist with efficient operation of the system. The system consists of a vertical discharge stack to which are connected soil or waste appliances via their own branch pipe. In order to prevent the penetration of sewer smells into the premises a water trap is used (Fig. 3.90).

If a trap seal is lost there is a direct pathway for sewer or drain odours to enter the building. The water seal depth is 50 mm for WCs and 75 mm for all other appliances. There are three principal ways (Fig. 3.91) in which the seal may be lost:

- self-siphonage
- induced siphonage
- back pressure

In addition to the main discharge stack a stub stack may be used. This allows connections from an unventilated stack (the

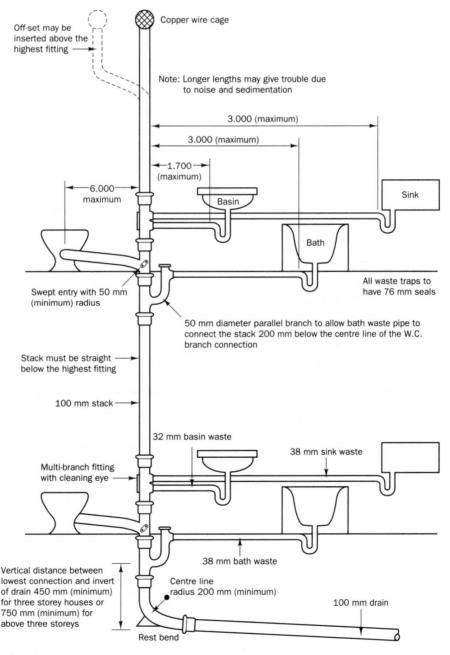

Figure 3.89 Single stack systems

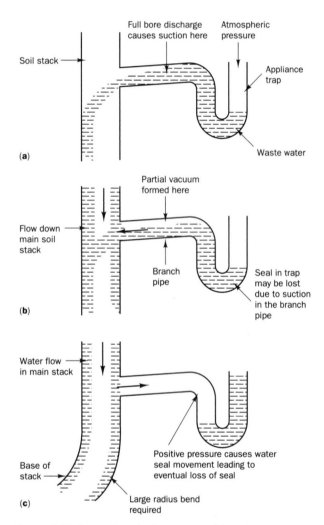

Figure 3.90 Types of traps: (a) P trap; (b) S trap; (c) bottle; (d) resealing; (e) running

Figure 3.91 Loss of seals from traps: (a) self-siphonage; (b) induced siphonage; (c) back pressure

stub stack) to a ventilated discharge stack or sewer. Conditions limit the connection of a branch to the stub stack to no more than 2 metres above the invert and a WC branch should be no more than 1.5 metres above the invert. The length of the unventilated stub stack should be no more than 6 metres for a single appliance or 12 metres for a group of appliances.

Below ground drainage systems

The system of drainage used will depend upon the locality and the requirements of the regional water undertaking. Two principal methods are used:

- separate system
- combined system

The separate system is commonly used. This collects surface water and foul water separately from a property and discharges them to separate sewers. The combined system collects foul water from appliances and surface water from paved and roofed areas by a single drain to a combined sewer (Fig. 3.92).

The design and installation of drainage systems should:

- be simple
- have few changes in direction and gradient
- sweep connections in the direction of flow

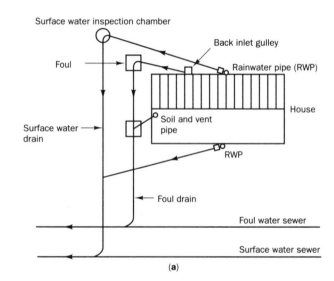

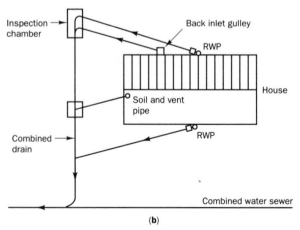

Figure 3.92 Drainage systems: (a) separate systems; (b) combined

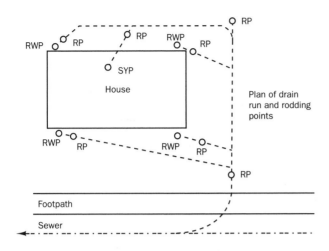

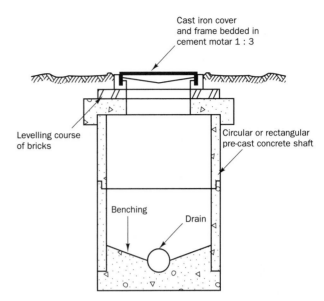

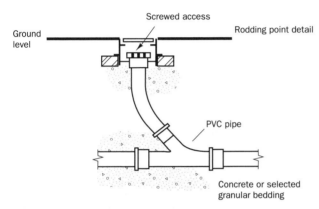

Figure 3.93 Typical rodding point system

Figure 3.94 Pre-cast concrete inspection chamber

- be ventilated at the head of the drain
- have pipes laid in straight lines
- minimise the effects of settlement
- use a self cleansing velocity of 0.75 m/s
- use impervious, strong and durable materials for pipes
- allow access for clearing blockages

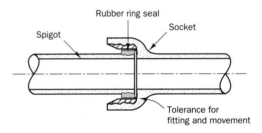

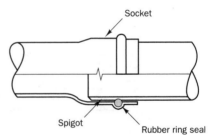

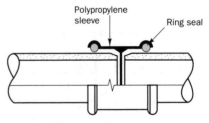

Figure 3.95 Flexible joints for pipelines

Table 3.18 Comparison of combined and separate drainage systems

Combined system	Separate system
1. Foul and surface water served by one drain and sewer which reduces cost	Two sets of drains are needed which increases installation costs
2. Risk of making wrong connection is eliminated	Possibility of making wrong connection exists
3. Surface water can be used to flush the foul water through the drain	Foul water drain relies on self cleansing gradient
4. Foul gases may bypass trap seals in rainwater gullies	No risk of polluting air through the surface water drain
5. Sewage disposal plant and treatment costs increase	Sewage disposal plant is smaller and treatment costs less
6. Pumping costs may increase but depends on location of the sewage treatment plant	Surface water may be disposed of locally thereby reducing the cost of any pumping

Pipes for drainage fall into two categories:

- rigid
- flexible

Rigid pipes are manufactured from vitrified clay, concrete, and grey iron. Flexible pipes are manufactured from uPVC. Joints used in pipelines should be appropriate to the materials being used. Flexible jointing methods are preferred to minimise the effect of ground movement through settlement (Fig. 3.95).

Flexible pipes should be laid at a minimum depth of 900 mm under any road and 600 mm in fields and gardens up

Types of access and their spacing are contained in the Approved Document of the Building Regulations. Two typical ones are rodding eyes (Fig. 3.93) and manholes (Fig. 3.94).

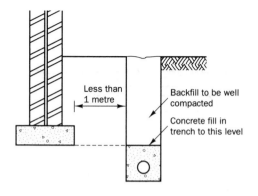

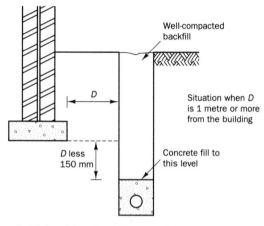

Figure 3.96 Provisions for drains near buildings

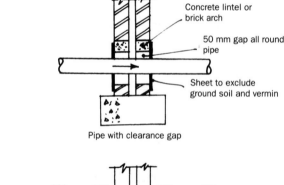

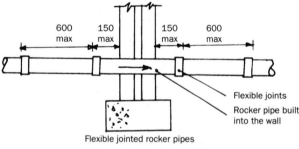

Figure 3.97 Drain passing through a wall

to a maximum depth of 10.00 m. Where depths are less than these then special protection of pipes will be required. Where drains pass alongside buildings the provisions shown in Fig. 3.96 should be observed. Figure 3.97 shows a method of accommodating a drain passing through a wall.

The choice of bedding and backfill to drains depends upon:

• the depth at which the pipes are laid

• their size and strength
• the nature of the ground excavated

Details of different bedding to drains can be found in the Building Regulations Approved Document.

Sanitary fittings for domestic premises

A variety of sanitary appliances can be installed to satisfy the range of special applications. Sanitary fittings can be viewed as:

• **soil fittings**
• **waste fittings**

Table 3.19 Sanitary fittings

Type	Fitting	Materials
Soil	WC	Vitreous glazed china
Waste	Wash-basin	Vitreous china Glazed fireclay Plastic
	Bath	Enamelled pressed steel Enamel on cast iron Fibreglass reinforced plastic
	Showers	Vitreous china Acrylic plastics Vitreous enamelled sheet steel
	Bidet	Plastic Vitreous china

Self-assessment tasks

1. State ways of disposing of surface water from a building.
2. Describe how trap seals may be lost.
3. What is a 'stub stack' and where is it used?
4. Describe and illustrate the differences between combined and separate drainage systems.
5. What are the advantages of using flexible pipe joints in foul drains?
6. Prepare a collage which illustrates the various sanitary fittings to be found in a college and the essential characteristics each must possess.

Space heating

Buildings are heated for various reasons but in domestic properties they are heated for the health and comfort of the occupants. An individual's satisfaction with the heating in a space is often a matter for personal preference, and may be related to the type of activity being undertaken, or the amount of clothing being worn. In domestic properties the acceptable temperature range lies between 19 °C and 23 °C.

Heating systems commonly used in domestic properties usually serve to modify the air temperature and not humidity. This does not matter too much as humans can feel comfortable in a wide humidity range. For normal comfort conditions this is between 30 and 65 per cent humidity.

Factors which influence heating installations

Apart from temperature and humidity the following will also have an influence:

- the orientation of the room or building
- use of the room
- volume of the room
- number of people in the room
- number and area of external walls and windows
- heat transmission levels through the structure
- temperature differences
- heat gain from appliances

In addition to the previous list the type of space heating installation will be a major influence. Systems may be direct or indirect. Direct systems are those where the energy purchased is consumed within the space to be heated (e.g. a gas or electric fire).

Indirect systems involve the energy purchased being consumed outside of the heated space and then transferred to the space for distribution.

Table 3.20 Fittings and components of a wet heating system

Component/Fitting	Purpose
Boiler	A source of heating water
Pipework	Used to carry the heated water
Control valves	Used to isolate or regulate flow
Motorised valves	Valves with electric motors allowing automatic regulation of flow
Radiators	Hollow metal heat emitters
Programmer	A control device, modern ones allow a variety of time settings throughout the day and week
Thermostat	Controls air temperature in a heated building or space

Hot water heating

These systems use water which is heated in a boiler which then transfers to the heating circuit called a primary circuit. Water is moved around the system by using a circulating pump. The use of the pump:

- allows efficient and effective control
- reduces pipe sizes
- provides quicker heat-up times
- increases flexiblity of the installation

The pump provides a velocity of up to 3 m/s and reduces the warm-up time. Pipework layouts can be adapted to suit individual building styles and are normally installed in a two-pipe ring circuit (Fig. 3.98). The two-pipe system uses two pipes called flow and return to feed the radiators. This system gives a balanced flow temperature to the radiators. Other forms of heat emitters can also be used, these being:

- fanned convectors
- skirting convectors

Water or 'wet' systems have an advantage in that the distribution takes place through pipes which are small in diameter and are easily routed around the property from downstairs to upstairs and even through timber floors. The pipes are therefore easily accommodated and may be completely out of sight or situated in positions which allow the pipes to be unobtrusive.

A variant of the system is the microbore installation. It is similar to the previous system except that the pipework is of a smaller diameter which allows the pipework to be accommodated more easily in both new and renovation installations. It also uses a manifold which the flow and return pass through (Fig. 3.99).

Self-assessment tasks

1. State the air temperatures which humans accept as comfortable.
2. Describe with examples factors that would influence the design of a space heating system.
3. How does micro-bore heating differ from small bore?
4. Describe the controls used with a wet heating system.
5. Using the controls identified in 4 list 3 different manufacturers of each control.

Installation and distribution of primary services

Buildings serving different purposes will require variations in the number, location and distribution of electrical, mechanical and sanitary services. In domestic premises this is not normally a problem, but where buildings become service intensive it is important that the design team collaborate at an early stage to minimises problems.

Access to services is needed to allow for:

- installation and minimal disruption to other activities
- inspection and maintenance
- isolation of services
- testing and commissioning
- alteration and replacement.

Distribution of services

Building services must be distributed about the building in an orderly and logical method. Two broad classifications for achieving this are structural and non-structural methods of accommodation.

Structural accommodation can be influential in the design of the building as it represents ducts and recesses accommodated within the fabric of the building. **Non-structural** methods involves attaching services to the building structure and then enclosing them with a secondary element such as a suspended ceiling or raised floor.

Inevitably the distribution of the services around the building will involve vertical and horizontal ducts. Additionally, from the main horizontal duct there will need to be lateral ducts to enable further distribution.

Services requiring consideration

Where air is used for heating or ventilation, **air trunking** is required. Trunking is normally the largest of the services and

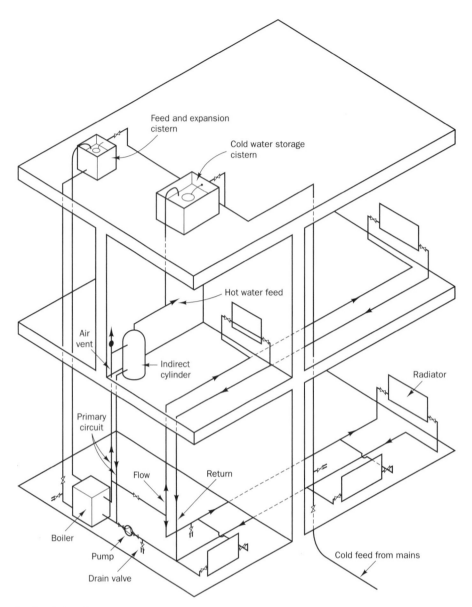

Feed and expansion cistern

Cold water storage cistern

Hot water feed

Air vent

Indirect cylinder

Radiator

Primary circuit

Flow

Return

Boiler

Pump

Drain valve

Cold feed from mains

Figure 3.98 Two-pipe heating system

therefore it will influence not only building design but also the layout of other services.

Waste and soil installations form a major limiting factor as their efficient operation requires them to be in vertical or horizontal planes (laid to falls). **Space heating** will require predetermined routes but often their path can be varied more easily than waste and soil pipes.

Hot and cold water supplies often use smaller pipework than heating. This enables their distribution to be flexible. As the operation of water supplies is not too dependent upon straight pipework they are able to be routed above or below other services. **Gas services** likewise present few problems.

Cable services for **electricity** and **communications** are very small and are often disposed of within wall or floor structures or in suspended ceiling voids. Alternatively plastic or metal trunkings are used.

Closely associated with services distribution is their location and treatment with respect to preventing spread of fire or enabling sound transmission from one area to another. Statutory requirements will also need to be observed.

Access methods

The method by which access is gained must relate to the considerations outlined previously. The principal techniques used are:

- ducts
- raised floors
- suspended ceilings
- access panels

Ducts

In buildings requiring extensive servicing it would be normal to distribute the pipe and cable services around the floor plan of the building from which vertical ducts will rise through the building. Ground floor ducts may be walkway or crawlway (Fig. 3.100). Access to walkway ducts will normally be from the boiler room or, as with crawlways, through lift-off covers strategically placed.

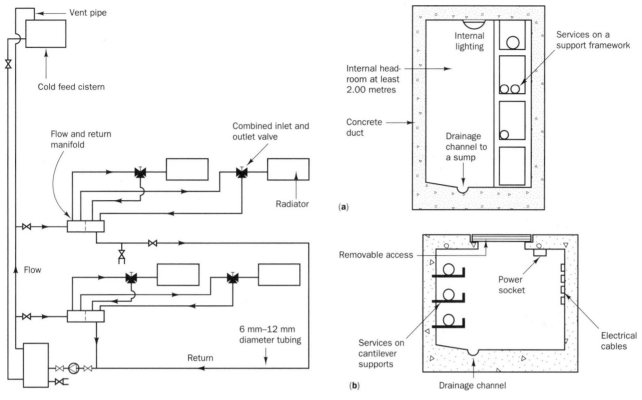

Figure 3.99 Micro-bore heating system

Figure 3.100 Ground floor ducts: (a) walkway; (b) crawlway

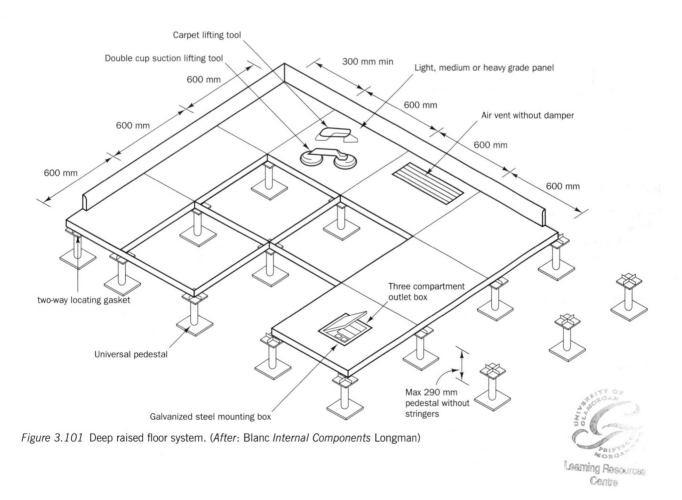

Figure 3.101 Deep raised floor system. (*After*: Blanc *Internal Components* Longman)

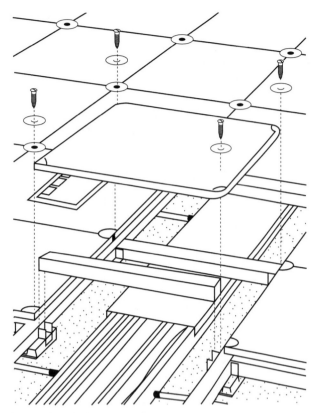

Figure 3.102 Shallow battened floor system. (*After*: Blanc *Internal Components* Longman)

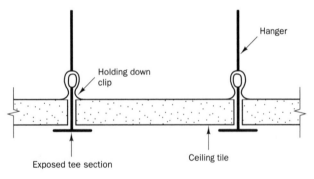

Figure 3.103 Exposed grid ceiling

Raised floors

Raised access floors originally provided accommodation for cabling associated with computer installations or the electronic office. Their use has now been extended to general office areas and other serviced spaces. The construction of the floor can vary in height and means of access. Access is possible through every floor panel or at designated points of access. Deeper voids will allow a crawlway provision (Fig. 3.101) whereas shallow voids will allow only surface access (Fig. 3.102).

Suspended ceilings

The lateral distribution of services from vertical ducts can be achieved by either raised floors or suspended ceilings. The common ceiling types are jointless, and frame and panel. Jointless ceilings will require special access provision for services and would not be suitable for heavily surfaced ceiling voids. It must be remembered that suspended ceilings are not

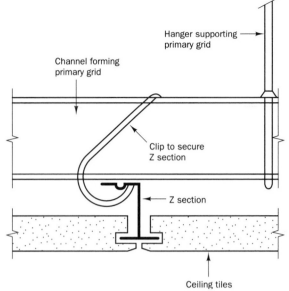

Figure 3.104 Concealed ceiling grid

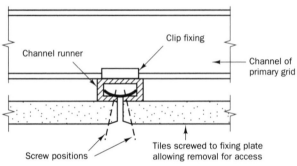

Figure 3.105 Access to concealed grid ceiling

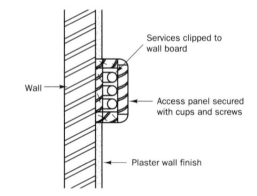

Figure 3.106 Surface duct

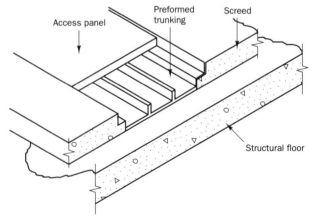

Figure 3.107 Floor duct

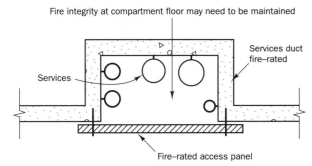

Figure 3.108 External wall duct

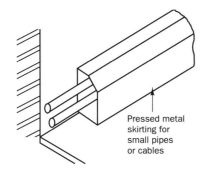

Figure 3.109 Skirting duct

a substitute for correct service support in ceiling voids. They are designed to carry only their own weight plus lighting and associated services.

Frame and panel systems consist of panels which may be laid directly into an exposed grid (Fig. 3.103) or where the grid is semi- or fully concealed (Fig. 3.104). The exposed grid will allow panels to be removed across the ceiling area but will not necessarily permit installation or removal of services without removal of some of the framework. Access to the concealed grid can be by screwed access (Fig. 3.105).

Access panels

Ducts that are to contain small pipes or cables are often provided on the surface of a wall (Fig. 3.106) or within the floor (Fig. 3.107) or wall construction (Fig. 3.108). Services may also be concealed in skirtings around the perimeter of a room (Fig. 3.109). In all cases screwed access panels will be used.

Self-assessment tasks

1. Why must access to services be considered at the design stage of a building?
2. Explain why some services will take priority over others.
3. Describe how suspended ceilings will allow access to services.
4. Sketch details of how access can be gained to services in raised floors.
5. Explain how access to crawlways may be achieved.

3.4 External works for a residential development

Residential properties are built in a variety of locations. These may be:

- rural isolated locations for a single property
- undeveloped land for a number of properties
- on land where previous properties have been demolished
- on infill locations behind existing properties

This section is concerned with a plot of land which has been previously undeveloped and which is to have a residential development of around fifteen detached properties. Such a development would be termed a small estate.

All building sites are different and therefore the ideal one does not exist. The building contractor will as part of the contract have to prepare the site to:

- enable detailed setting out to take place
- modify the ground contours to remove excessive undulations
- provide for services to properties and the access roads
- provide retaining structures to accommodate severe changes in ground levels
- allow access about the estate both during construction and when occupied

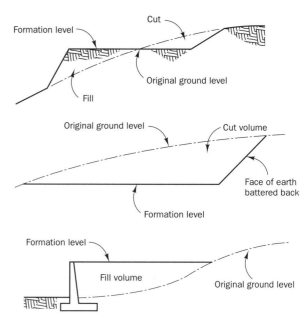

Figure 3.110 Cut and fill techniques

Forms of earthwork for a residential development

Prior to any earthworks being carried out site clearance will have been necessary. On undeveloped or 'greenfield' sites this would have involved:

- identification, protection and isolation of existing services
- removal of bushes, hedgerows or trees
- formation of a site access point
- erection of hoardings around the site to restrict entry
- stripping of topsoil and storing it on site or loading and carting it away
- retention and protection of any special features, e.g. trees, fencing

Cut and fill

When sites have been stripped of their topsoil and the contours are still undulating it may be necessary to use cut and fill techniques. The purpose of this is to move earth from one position to another thus changing the ground levels to:

- enable setting out of the buildings
- provide particular landscaping features
- provide level areas for the construction of buildings
- assist with the construction of roads

Three techniques (Fig. 3.110) may be used to modify ground levels, these are:

- cut
- cut and fill
- fill

Cut levels an area and has the advantage of leaving stable ground; cut and fill uses the cut part of the ground to fill the low areas; and fill involves using suitable material from elsewhere for low areas. It should be noted that fill areas can be subject to settlement and if they are to be used to support a structure may need thorough compaction, or the careful choice of a foundation type to resist settlement.

Landscaping

The importance of landscaping should not be under estimated. It assists in providing both a present and future environment for mankind to enjoy. Landscaping is commonly described in categories of hard or soft. Hard is associated with surface finishes, fencing, bollards, seating, lamp posts and flag staffs for example. Soft landscaping covers items such as grassed areas, trees, shrubs and plants.

Grassed areas apart from providing ground cover also have the following advantages:

- a variety of textures
- restful colours
- resilience to wear
- absorbs sound
- cheap to establish

Correct preparation for grassed areas is necessary if the grass is to grow and flourish. Care must be given to:

- levelling and gradients of the ground
- surface and subsoil drainage
- correct selection of seed or turf to match the use
- type and regularity of maintenance

Trees take time to reach maturity but correct choice and siting initially will bring benefits such as:

- providing a decorative backdrop or feature
- providing screens from the wind or sun
- filtering dust from the air
- stabilising the ground

Every encouragement is given to attempt to retain existing trees in new building schemes. Where this is carried out their protection during the construction phase is important. In some instances the local authority may make a tree preservation order which may apply to a single tree, groups of trees or woodlands. Trees in conservation areas are also protected.

Factors which will affect trees and their growth are:

- the finished ground levels
- position of services and drains
- distance to adjacent buildings, roads and footpaths
- soil and climatic conditions

While trees and shrubs may assist in stabilising the ground they may not be suitable where steeply sloping banks are produced. Bio-engineering techniques are being used more and consist of:

- filter fabrics and meshes which prevent rotational slipping
- using degradable meshes to enable plant colonisation to take place
- mats of pre-seeded material being pinned to a bank while growth occurs
- natural materials or plants which root quickly and stabilise the earth

Trenches

Trenches are a form of excavation used to:

- accommodate building services
- provide certain types of land drainage
- enable building foundations to be formed

Their sides may be vertical or battered (Fig. 3.111). Vertical sided trenches should not be assumed to be safe or self supporting. Ground conditions can change according to weather conditions and ground water content in the soil. Collapse of trenches is a real possibility and poses a threat to the safety of operatives.

Safe working conditions in excavations are controlled by the Construction (General Provisions) Regulations 1961 and The Construction (Health, Safety and Welfare) Regulations 1996 and under the general requirements of the Health and Safety at Work etc. Act 1974.

Unless the sides of the trench are battered back to a safe angle the trench sides will need to be supported. Types of support will vary and will need to take into account the following:

- ground conditions
- excavation depth
- type and size of the excavation
- ground water level
- proximity of roads, buildings and buried services

Trench support for firm soils is shown in Fig. 3.112, and for loose wet soil in Fig. 3.113. Further guidance relating to excavations can be found in British Standard 8000: 1989, 'Workmanship on Building Sites' Parts 1 and 14.

Where excavations exceed more than 2 metres in depth they should either be covered or have suitable safety barriers erected. Excavations should be inspected by a

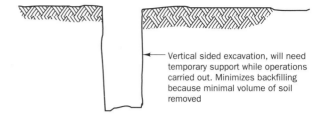

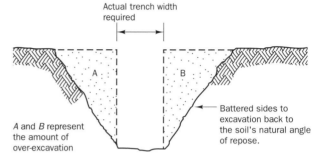

Figure 3.111 Trench excavations

Vertical sided excavation, will need temporary support while operations carried out. Minimizes backfilling because minimal volume of soil removed

Actual trench width required

A and B represent the amount of over-excavation

Battered sides to excavation back to the soil's natural angle of repose.

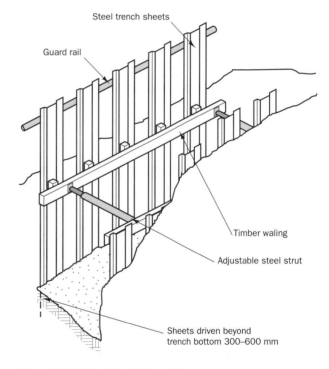

Steel trench sheets

Guard rail

Timber waling

Adjustable steel strut

Sheets driven beyond trench bottom 300–600 mm

Figure 3.112 Trenches in firm soil

competent and experienced person before the start of every shift, and every day that people are working in the excavation. A thorough examination should be carried out every seven days, with a record of the inspection being kept.

Services installations

A small estate will require the following services to be provided:

- gas
- water

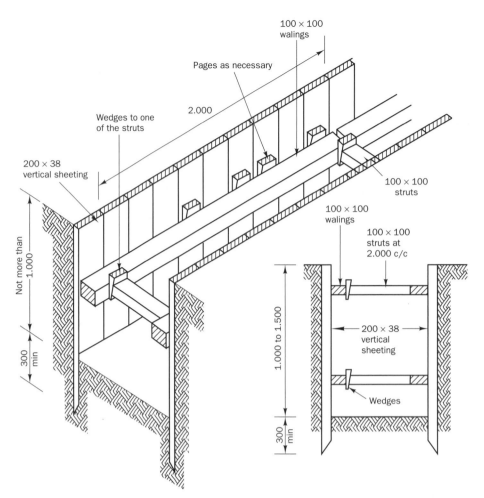

Figure 3.113 Trenches in loose wet soil

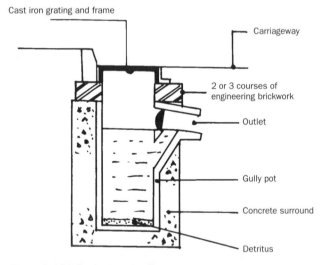

Figure 3.114 Typical road gully

- electricity
- drainage
- telecommunications
- external lighting

The first four of these have been discussed in section 3.3. Here highway drainage, telecommunications and lighting will be considered.

Highway drainage

The aim of road drainage is to create a route along which water from the highway will flow to a suitable discharge point. A ditch at ground level may be suitable or pipes below ground level are alternatives. Suitable discharge points may be another drain or sewer, watercourse or soakaway. Pipes are preferred to carry the water from the road as they are most efficient.

Surface water is collected into channels and discharged by gullies (Fig. 3.114) to a sewer system or other discharge point. A small sump in the gully is used to collect detritus, thus preventing the drainage system from becoming blocked at a later time. The gully may be situated either flush with the surface of the road or so water enters it from an inlet in the kerb.

Gullies are spaced at intervals along the road which is constructed so that water will flow to them. Transverse gradients of between 1:35 and 1:50 and longitudinal gradients of 1:100 to 1:200 are used, the gradient being dependent upon the type of surfacing material. Each gully drains between 190 m² to 220 m² of road surface.

Telecommunications

In rural areas overhead lines are still used but for urban areas cables are laid below ground in a network of ducting, with access chambers at suitable intervals. Often they are routed under the footway up to 600 mm in depth. This then avoids other services of gas, water and electricity in particular. Connection to individual properties then occurs. Where the cable passes in to the building a protection sleeve is used.

External lighting

Guidance on the general principles of road lighting are contained in British Standard 5489 Part 1. Recommendations for lighting access roads, pedestrian areas and residential roads are covered in Part 3. General criteria for lighting are to provide good visual conditions by achieving:

- appropriate levels of surface luminance
- uniformity of luminance distribution
- minimum glare from lighting equipment

In residential areas the lights may be at heights ranging from 5 m to 8 m and the discharge lamps from 35 W to 150 W. It is normal to stagger the position of lighting columns but usual to find them situated at road junctions in particular.

For small estates and other residential developments lighting should also consider the safety and security of pedestrians. Lighting is required especially around:

- steps
- ramps
- low walls
- pedestrian crossings

Operations and plant used to form earthworks

There are various tasks related to external works which can be carried out by using mechanical plant (Table 3.21). A machine used on small estate developments is the combined loader and backacter. Its use produces economic advantages because:

- output is increased
- work in inclement weather is possible
- movement of greater volumes is possible
- mechanical plant is reliable

Table 3.21 Operations performed in external works

Operation performed	Activity
Excavation	Forming ditches
	Pits for gullies
	Pits for tree planting
	Service trenches
	Stripping topsoil
	Trenches for retaining walls
Earth and material moving	Landscaping
	Loading excavated earth
	Preparing formation levels for roads
	Spreading granular material or hardcore
	Trench backfilling
	Trimming of excavations

Functions and structure of gravity retaining walls

Retaining walls are used in external works on sites to:

- allow terracing of sloping sites
- retain earth to allow construction of a footway, roads or parking bay

The retaining wall has various forces acting on it (Fig. 3.115), which require it to fulfil the following functions:

- resist the horizontal pressure from the retained earth
- resist ground water pressure in the retained earth

In order to achieve this effectively the wall must be constructed in such a way that it resists:

- rotational failure
- failure by sliding
- slip circle failure
- structural failure (Fig. 3.116)

The construction and form of a retaining wall will be dependent upon the:

- height of the wall
- type of material being retained
- ground conditions
- material to be used to construct the wall
- access
- aesthetics

Up to 3 m in height it is possible to use a mass or gravity retaining wall. This can either be constructed from brickwork

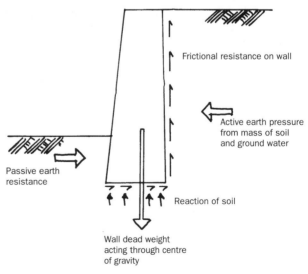

Figure 3.115 Forces acting on a retaining wall

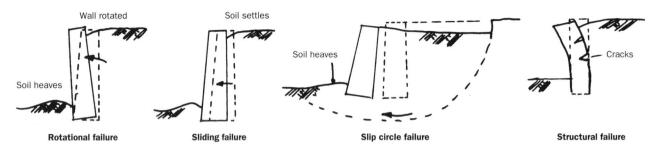

Figure 3.116 Types of failure on a retaining wall

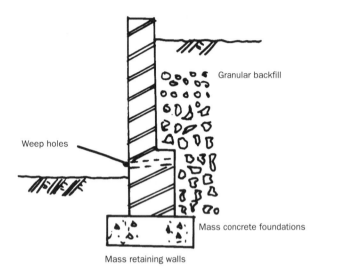

Figure 3.117 Gravity retaining walls

or concrete (Fig. 3.117). They depend upon their own mass together with the frictional resistance against the soil for their stability. The design based on the gravity principle allows for no tension to exist, nor for small tensile stresses to develop.

They may be constructed with their faces either inclined or stepped. The weight of the soil acting on the steps can increase stability due to the additional downward force. The width of the base of a mass retaining wall generally has to be a quarter to a half of the depth of the earth being retained, depending upon the soil properties, to prevent tension being developed.

It will be necessary to provide some form of drainage to release water from the rear of the retaining wall and also to reduce the hydrostatic pressure. This can be achieved by either providing weepholes through the wall at approximately 900 mm centres, or by including a positive drain at the base of the wall with porous filtering material above.

Self-assessment tasks

1. Briefly describe why retaining walls are used.
2. Explain how a gravity retaining wall maintains its stability.
3. Describe how ground water pressure at the rear of a retaining wall may be controlled.

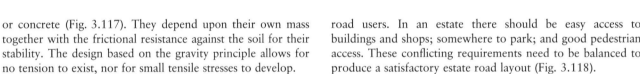

Forms of roadworks for an estate road

The needs of drivers and pedestrians associated with small residential estates will differ from the requirements of major road users. In an estate there should be easy access to buildings and shops; somewhere to park; and good pedestrian access. These conflicting requirements need to be balanced to produce a satisfactory estate road layout (Fig. 3.118).

Roads

Roads on estates are a mixture of collector roads, cul-de-sacs and hard standing. Collector roads permit vehicles to move within a neighbourhood and provide some access to individual properties. Cul-de-sacs or access roads tend to be used exclusively for reaching individual properties. Hard standings are areas where parking is permitted.

Figure 3.118 Typical estate road

The surface of the road and its associated construction is known as a pavement. A pavement is a surface which is intended for traffic and where the soil is protected by an overlay of imported or treated material. This protection is provided by layers of selected materials which transfer and distribute traffic loads to the ground, the objective being to limit the stresses in the ground.

Pavements may be described as:

● rigid
● flexible
● semi-rigid

Rigid pavement construction

Rigid pavements consist of a surface slab, a sub-base and the sub-grade (Fig. 3.119). The surface slab of high quality concrete forms a structural plate, and may be reinforced with steel bars or mesh. A general characteristic is that the slab is able to span over minor irregularities in the sub-grade.

When a sub-grade is incapable of giving consistent support to a slab because of frost action, movement or poor drainage a sub-base is used. This is formed from either granular material or a weak mix of concrete with a thin sealing layer of bitumen to prevent water absorption.

The sub-grade is the natural foundation and is created by material having been excavated from above it, or by overlaying imported fill which will need to be adequately compacted.

The concrete surface slab will be subject to temperature changes which can cause expansion and contraction as well as warping. If the slab is continuous then it is likely that it will fail before the end of its useful life. Joints are therefore introduced to minimise movement. Contraction and expansion joints are shown at Fig. 3.120.

Flexible pavement construction

A characteristic of a flexible pavement is that it is able to maintain its structural integrity when small vertical movements take place at the surface. Principal components of the flexible pavement are the surfacing course, the roadbase and sub-base (Fig. 3.121).

The surfacing course provides a running surface for vehicular traffic and prevents surface water penetrating into the pavement. The surfacing is in two layers which can be laid to fine tolerances to provide:

● surface regularity
● acceptable surface texture
● durability
● flexibility

The roadbase contributes to the strength of the pavement and is able to distribute wheel loads through to the sub-grade. The roadbase may be formed by using cement or bitumen bound

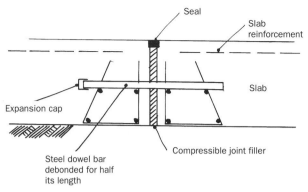

Expansion joint

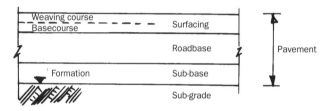

Contraction joint

Figure 3.120 Typical expansion and contraction joints

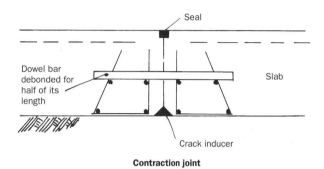

Figure 3.121 Flexible pavement construction

materials, stabilised soil or carefully graded granular material.

The capping and sub-base are the pavement foundation. They provide a working platform which supports the roadbase, and insulate the sub-base against the weather.

Semi-rigid pavement construction

Block paving provides a very strong flexible surface which is able to support concentrated loads without undue deformation taking place. The blocks are laid in a course one block deep in graded sand within a firm edge restraint on a prepared foundation. The blocks are compacted into the sand with very thin joints between them. This creates strong frictional forces between the blocks. It also assists the deeper blocks in resisting eccentric loading.

The block pavings are made from concrete or clay ware and are 100 mm wide, 200 mm long, and between 65 mm and 100 mm thick. They may have straight or profiled interlocking edges and are available in a variety of colours to allow patterns to be formed.

This form of paving is ideally suited to situations where the traffic speed is unlikely to exceed 30 mph. They are used in residential developments for their attractive surface finish where large areas can be broken up to relieve visual monotony. Common uses in estate work are for:

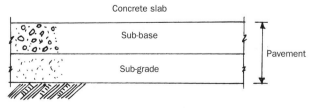

Figure 3.119 Rigid pavement construction

- access roads
- cul-de-sacs
- hard standings
- driveways

Pavings

Small element paving is not a recent innovation. In historic times materials such as cobbles, wood blocks and granite setts were used. Cobbles are used today but their use is limited to decorative applications.

Granite setts are still used, being extremely durable and able to take heavy wear. They are used in pedestrian areas and can be laid to produce a variety of patterns. They may be functional as a footway, or for decorative or ornamental use.

Thick pre-cast paving slabs are available which are dimensionally co-ordinated with block paving which has been described previously. The slabs are 400 mm square and can be interlocking. They are used in light traffic areas, not where there is continuous use by vehicles.

Hard standings

On small estates it is essential that parking is provided for residents and visitors. In some cases on-street parking may be sufficient, but where projections suggest it will not meet the demand highway authorities will produce specific requirements for a development. The use of hard standings and parking bays is often required. These are constructed in appropriate places using materials and forms of construction similar to footways and semi-rigid road construction.

Footways

Footways are routes associated with carriageways and provided for pedestrians being either adjacent or related by a verge. Their widths vary between 1.35 metres to 3.00 metres. Footpaths are routes away from traffic routes. Materials used for footways and footpaths are:

- block pavings
- stone paving
- concrete paving slabs
- granite setts
- tarmacadam over a sub-grade
- unreinforced concrete slab

In general footpaths and footways should:

- satisfy pedestrian walking lines
- be of sufficient width and well lit
- keep gradients to a minimum, over 8% is difficult for wheelchair users
- use a minimal number of steps
- have dropped kerbs at road crossings

Self-assessment tasks

1. Describe the difference between a rigid and a flexible pavement.
2. Sketch a detail of a contraction and an expansion joint for a rigid pavement.
3. Define 'footpath' and 'footway'.
4. State the general requirements for a footpath.

Construction Operations

Jeff Attfield

This unit introduces students to the processes involved in construction operations and the associated legislation necessary to ensure the health, safety and welfare of **operatives** and members of the public within the construction industry.

Any construction operation will contain an element of risk with potentially dangerous situations arising from time to time. To reduce this risk the student must be aware that a neat and tidy site/workshop is a major factor in reducing accidents and creating a good working environment for all operatives.

The student will also be shown how construction elements can be measured and costed to provide the basis for quotations and to let the builders know the amount of materials and operatives to use for the construction process.

Craft operations will be investigated. Some students will be lucky enough to practise their own personal practical skills in one or more of the trade craft workshops, while others may not be so fortunate, having to gain their knowledge through observation of practical activities.

The key areas covered by this chapter are:

- Health and safety requirements
- Measuring and costing the quantities of materials
- Planning construction operations for a building
- Basic craft operations

4.1 Health and safety requirements within the workplace

Topics covered in this section are:

- The purpose of health and safety legislation
- Duties under the Health and Safety at Work etc. Act
- Potentially dangerous situations within the workplace
- Methods of managing safety within the workplace

The purpose of health and safety legislation

Any construction works, whether they be minor works, such as house extensions, or large multi-million pound projects, such as shopping centres, are bound by law to provide for the health and safety of all operatives and the general public alike.

It is quite evident from the existing legislation that not only must the **employer** (the General Contractor) provide and maintain a safe and healthy working environment but equally so must the **employee** (Site Operative) or **self employed** Site Operative. All three parties can be prosecuted for their 'own acts or omissions'. It is important to understand that a self-employed person has the same duties as the employer, i.e. must not endanger themselves or others.

To explain this further, an **act** is something you have done, and an **omission** is something you have failed to do but should have done. In this context acts and omissions are specifically related to a person's duties in law, in which they have a duty of care not only to themselves but also to their fellow workers. This law is governed by Acts of Parliament known as the Factories Act 1961 and the Health and Safety at Work etc. Act 1974. These are general Acts to cover all types of work situations and practices and are not totally geared to the construction industry. The main substance stated in them provides the basis for other legislation which is specific to the construction industry and will be mentioned later.

Legislation within the construction industry is great in quantity and the student is only expected at this stage of study to have sufficient understanding to:

- describe the purpose and relevance of health and safety legislation
- identify duties under the Factories Act 1961 and the Health and Safety at Work etc. Act 1974
- identify potentially dangerous situations within the workplace
- describe organisational and physical protection used in the workplace
- identify methods of personal protection used in the workplace

In the industry each firm employing five or more persons must by law prepare a written Health and Safety Policy and ensure that their employees' attention is drawn to it, i.e. bringing it to their notice.

In large construction companies professional Safety Officers are employed to ensure compliance with the regulations; to promote the safe conduct of the work generally and to ensure that their sites are safe and do not infringe current legislation. Smaller companies who do not have sufficient work to employ a full-time safety officer can use the services of Safety Consultants who employ professional safety officers on a full-time basis.

Legislation for the health and safety of site or workshop operatives and members of the public is covered in the following:

- Construction (Health, Safety and Welfare) Regulations 1996
- Construction (Lifting Operations) Regulations 1961, as amended 1992 (to be revoked in 1997)
- Construction (Head Protection) Regulations 1989, as amended 1992
- Control of Substances Hazardous to Health Regulations 1988, amended 1993
- Health and Safety (First Aid) Regulations 1981

The legislation mentioned here is by no means the complete legislation for the construction industry. All you need to know at this stage in your studies is the areas mentioned above and to what they apply.

A greater depth of coverage of legislation affecting the industry will be required in the GNVQ Advanced Unit 6 Resource Management, which can be found in *Advanced GNVQ Construction and the Built Environment*, published by Addison Wesley Longman.

Health and Safety at Work etc. Act 1974

There are some 85 sections within the Health and Safety at Work etc. Act, split into four parts and ten schedules covering the following areas:

Part I: Health, safety and welfare in connection with work, and control of dangerous substances and certain emissions into the atmosphere.

- Section 1 is the preliminary
- Sections 2 to 9 – general duties
- Sections 10 to 14 – the Health and Safety Commission and the Health and Safety Executive
- Sections 15 to 17 – health and safety regulations and approved codes of practice
- Sections 18 to 26 – enforcement of regulations
- Sections 27 to 28 – obtaining and disclosure of information
- Sections 29 to 32 – special provisions to do with agriculture
- Sections 33 to 42 – types of offences and their administration
- Section 43 – financial provisions
- Sections 44 to 54 – miscellaneous and supplementary regulations

Part II: The Employment Medical Advisory Service – (Sections 55 to 60).

Part III: Building Regulations and Amendment of Building (Scotland) Act 1959 – (Sections 61 to 76).

Part IV: Miscellaneous and general – (Sections 77 to 85). Schedules 1 to 10.

Self-assessment tasks

Using a copy of Health and Safety at Work etc. Act 1974:

1. Make a list of the duties required of each person towards health and safety (Sections 2 to 9).
2. Make a list stating the powers of inspectors (can be found in enforcement sections).

The enforcement of safety, health and welfare legislation within the construction industry is the responsibility of the Health and Safety Executive (HSE). The HSE is a government department which employs inspectors (factory inspectors) to visit workplaces and/or construction sites to ensure that the regulations are not being contravened.

In some instances a Local Authority may employ Inspectors but only when authorised by the Secretary of State for Employment.

Powers of inspectors of health and safety

In 1983 the Factory Inspectorate celebrated 150 years of existence. The Factory Inspectorate is currently known as the Health and Safety Executive. Factory Inspectors have responsibility and authority (in all the current Acts) to administer and prosecute anyone contravening safety, health and welfare legislation in a majority of construction operations.

The main statutory powers are given in the Health and Safety at Work etc. Act 1974, Sections 18 to 26, and 39.

'It is the duty of the Health and Safety Executive to make adequate arrangements for the enforcement of the relevant statutory provisions.'

The Factory Inspector has powers to:

- Stop any works not complying with the Regulations until the infringement has been corrected. This is carried out by the issuing of a 'Prohibition Notice' preventing any further works by closing down all or part of the site. The closures can vary from a few hours to several days, and could be very expensive for a contractor in the form of lost production and operatives' time.
- Enter, examine and investigate at any reasonable time or at any time in a dangerous situation.
- Require any premises (or part) or anything in premises within his/her field of responsibility to be left undisturbed.
- Cause any article or substance in the premises to be dismantled or tested.
- Take possession of and detain the article or substance long enough for examination or for production in legal proceedings.
- Require a person to give information and sign a declaration.
- Take measurements, photographs and samples and to require production of relevant documents.

- Prosecute in a magistrates' court and issue Improvement and Prohibition Notices.
- Seize or render harmless any article or substance which is a cause of imminent danger or serious personal injury.

Potentially dangerous situations

To understand why we need all this legislation in the construction industry, we need to consider what is a potentially dangerous situation. Any one or a combination of the following are potentially dangerous:

- falling
- mangling
- treading on sharp objects
- striking against objects
- struck by falling objects
- machinery
- hand tools
- moving plant
- electricity on site
- noise
- hazardous material
- unsafe handling of solvents
- failure to use appropriate PPE (Personal Protective Equipment)

By now you are probably wondering why you have decided to enter such a potentially dangerous industry. Let us put this information into perspective. Every time you cross a road you put your life in danger with traffic passing in both directions (potentially dangerous situations) *but* if you observe the simple rules of the crossing code this can be carried out in complete safety. The same applies in the construction industry; providing you know the dangerous situations and move around observing the risks and knowing your safety legislation, you will be safe. It is only when operatives take risks that potential danger lurks.

Accident, disease and dangerous occurrences notification

Legislation applicable

- Health and Safety at Work etc. Act 1974
- Reporting of Injuries, Diseases and Dangerous Occurrences Regulations 1996 (RIDDOR)

Accidents in general

Notification of accidents in the workplace falls into three categories:

- fatal accidents
- major injury accidents
- all injuries resulting from accidents at work which cause incapacity for more than 3 days

If an operative has an accident in the workplace, and is off work (incapacity for work) as a direct consequence of that accident for three days, then by law the accident has to be reported to the Health and Safety Executive on a Form 2508. This may well result in a Factory Inspector making a visit to the scene to carry out an investigation using all or part of his/her powers to come to a ruling as to the cause. Prosecution under the Health and Safety at Work etc. Act may take place as a result.

Fatal accidents

Fatal accidents have to be reported if they arise out of or in connection with work whether the person who dies is employed or not. For instance it could have been a member of the public in the street killed by a scaffold collapse. It could also be a young child killed while trespassing/playing on site after hours and falling into an open trench.

A further duty exists to provide a written report of an accident which results in a death within one year of the date of the accident. It should be submitted to the Health and Safety Executive. This requirement exists even if the initial injury was either reported initially as non-fatal or not originally reported.

Major injury

Accidents resulting in major injury will also have to be reported. Major injury is defined as:

- fracture of the skull, spine or pelvis
- fracture of any bone in the arm or wrist, but not a bone in the foot
- amputation of a hand or foot, or a finger, thumb or toe, or any part thereof if the joint or bone is completely severed
- the loss of sight of an eye, a penetrating injury to an eye, or a chemical or hot metal burn to an eye
- either injury (including burns) requiring immediate medical treatment, or loss of consciousness, resulting in either case from an electric shock from any electrical circuit or apparatus, whether or not due to direct contact
- loss of consciousness resulting from lack of oxygen
- either acute illness requiring treatment, or loss of consciousness, resulting in either case by inhalation, ingestion or absorption through the skin
- acute illness requiring medical treatment where there is a reason to believe that this resulted from exposure to a pathogen or infected material
- any other injury which results in the person injured being admitted into hospital for more than 24 hours

Three day accidents

Accidents causing more than three days' incapacity for work are reportable. An HSE Form 2508 has to be completed and forwarded to the appropriate enforcing authority within seven days of the accident. However, no immediate notification is required.

Note that in counting the three days, the day of the accident is not included. However, any days which would not normally be working days (such as Saturday and Sunday) are included. 'Incapacity for work' does not necessarily mean actual absence from work. Inability to perform work that the person might reasonably be expected to do is equally 'incapacity for work'.

To give some idea of the potential dangerous situations, it is necessary to look at the statistics that exist.

Table 4.1 Accidents by activities

Activity	Percentage of reportable accidents
Transport	7%
Machinery	8%
Hand tools	8%
Struck by falling objects	9%
Striking against an object	12%
Handling	25%
Persons falling	26%
Other (miscellaneous)	5%

As you will see the most common type of accident is that of persons falling. Anyone working above 1.98 m must be protected from falling by the use of handrails and toeboards: the reason for this is that a body falling will rotate through 180 degrees over this distance thus causing it to land head first which is sufficient to cause fatal consequences.

A reference book showing how to manage health and safety on sites to prevent accidents is published by the Health and Safety Executive – *Health and Safety for Small Construction Sites*, ISBN 0 7176 0806 9.

The Construction (Health, Safety and Welfare) Regulations 1996

These regulations are aimed at securing the protection of people employed on building operations or work of engineering construction. They impose requirements as to the health, safety and welfare provisions on building, construction and engineering sites. They do not apply to any workplace on a building or construction site that is set aside for purposes other than construction work.

The term **construction work** includes the following applications:

- Construction, alteration, conversion, fitting out, renovation, repair
- Site clearance, exploration or preparation for an intended structure
- Assembly of prefabricated elements
- Removal of the whole or part of a structure
- Installation, commissioning or repair of mechanical, electrical, gas and other types of services

It is the duty of every employer whose employees are involved with construction work, and every self-employed person and every employee, to comply with the regulations that affect them or any person under their control. Every person at work should co-operate with the employer and report any defect that may endanger health and safety.

Detailed knowledge is not expected of a student of the regulations but a knowledge of the areas they cover is anticipated. The following is a summary of the areas where the regulations apply:

Safe places of work	Fresh air
Falls	Temperature and weather
Fragile materials	protection
Falling objects	Lighting
Stability of structures	Good order
Demolition or drowning	Plant and equipment
Explosives	Training
Excavations	Inspection
Reports	Exemption certificates
Prevention from drowning	Extension outside Great
Traffic routes	Britain
Doors and gates	Enforcement in respect of fire
Vehicles	Modifications
Prevention of risk from fire	Revocations
Emergency routes and exits	Fire detection and fire
Emergency procedures	fighting
Welfare facilities	Coffer dams and caissons

Inspections are an important part of these regulations and should be carried out by a competent person. Inspections are expected of:

● Working platforms, personal suspension equipment
● Any excavation which is supported and may be in a temporary state of weakness or collapse
● Coffer dams and caissons

After an inspection has been carried out a report is to be made within 24 hours. This must clearly describe the findings of the inspection to the person on whose behalf the inspection was carried out. The report, or a copy of it, should be kept on site for a period of three months from the date of the completion of all work on the project.

A **competent person** is defined as someone who has practical and theoretical knowledge and actual experience of what they are to examine, so as to enable them to detect errors, defects, faults or weaknesses, which it is the purpose of the examination or inspection to discover; and to assess the importance of such discovery.

The Construction (Head Protection) Regulations 1989

These regulations require the provision and wearing of suitable head protection on sites, for the safeguarding against head injuries of personnel involved with construction activities.

Suitable head protection means any head protection that is designed as far as is reasonably practicable to protect the wearer against any foreseeable risk or injury to their head. It must be suitable for the work activities being undertaken and must, with necessary adjustment, fit the wearer comfortably.

Employers must provide all employees with suitable head

protection where, due to the nature of the work or activity, they are liable to be at risk from head injuries. Self-employed people have an obligation to supply themselves with suitable head protection in similar circumstances. The head protection should be suitably maintained and replaced when necessary.

Head protection must comply with the requirements of The Personal Protective Equipment at Work Regulations 1992. Storage for head protection should be available when it is not in use.

Self-assessment tasks

Research the Head Protection Regulations to ascertain:

1. Who is responsible for providing head protection?
2. What duty of care has an employee over his/her head protection?
3. When should head protection be worn?
4. How does a person in charge of a site instigate the wearing of head protection?

The Construction (Design and Management) Regulations 1994

These regulations set out the way in which all building and construction work that is within the scope of the regulations should be designed and managed from the initial concept of a project to its final completion. The regulations apply to all notifiable construction work under the regulations. When demolition work is being carried out, all of the provisions of the regulations apply irrespective of the size of the project. It should also be noted that The Health and Safety at Work etc. Act and all other relevant statutory provisions will continue to apply.

The regulations impose duties on:

● the client
● designers
● developers
● planning supervisors
● principal contractors
● contractors

A key part of the regulations is the **health and safety plan**. If it is properly prepared and then properly used, compliance with it will improve the management of health and safety on site, and also reduce the number of accidents and incidents of ill health.

The health and safety plan consists of two parts:

● pre-tender stage
● construction phase

At the pre-tender stage the planning supervisor must ensure that a health and safety plan is prepared before any arrangements are made for a contractor to carry out or manage any of the work.

The **pre-tender health and safety plan** is supplied to persons who are being invited to tender for the work. It will give the contractors all the information they will need to demonstrate their competence, and it will also make them aware of the potential health and safety problems that may arise on site during the life of the project.

At the construction phase the planning supervisor passes on the health and safety plan to the principal contractor. It is then that person's responsibility to develop it into a working document that will assist in maintaining health and safety management of the project.

The planning officer has a resonsibility to ensure that a **health and safety file** is prepared for each building or structure that there is in the project. At the end of the project the file is given to the client and then remains with building or structure as a permanent record of how the project was designed and built.

The file must be kept by the client and the information contained within it made available to persons who need it before or when carrying out construction work at a later date.

The Control of Substances Hazardous to Health 1994

The construction industry uses a variety of substances that can represent a health risk to the people using or coming into contact with them. Examples include cement, solvents and resins. In addition, the processes used can generate hazardous substances – for example, fumes and toxic gases and dust from, say, grit blasting.

The purpose of these regulations is to safeguard the health of people using or coming into contact with substances that may be injurious to their health. A substance is regarded as hazardous to health if it is hazardous in the form in which it occurs in the workplace. Also, if anything causes an injury to the health of an employee then it should be regarded as a substance hazardous to health.

The main requirements of the regulations are:

● Employers must carry out an assessment of health risks created by work involving substances hazardous to health
● Employers must either prevent exposure of employees to substances hazardous to health or, where it is practicable, adequately control exposure
● Employers who provide any control measures are required to ensure that they are properly used or applied
● Adequate maintenance, examination and testing of control measures must be carried out
● Monitoring of the exposure to risk in the workplace should be undertaken
● Health surveillance must be carried out and recorded
● Information, instruction and training must be provided for persons who may be exposed to substances hazardous to health

A record of assessments should be kept in writing and be readily available. These should be reviewed regularly, with the maximum period not exceeding five years.

The Health and Safety (First Aid) Regulations 1981

These regulations place a general duty on employers to make or ensure there is made adequate first aid provision for their employees if they are injured or become ill at work. The employers must also inform the employees of the first aid provision made for them. Self-employed persons are also required to provide adequate equipment.

The regulations provide a flexible framework within which employers can develop effective first aid arrangements relative to their workplace. Four main factors will influence the level of first aid provision:

● number of employees
● the nature of the undertaking
● the size of the establishment and the distribution of the employees
● the location of the establishment and the locations to which employees go in the course of their work

First aid treatment should be recorded in an accident book. An additional book may be needed in order to record the use of first aid materials. All accidents causing any injury must be recorded and, where circumstances require it, reported to the enforcing authority.

Organisational methods of managing safety

Preventing accidents

Accidents can be prevented by controlling the activities and operations in the workplace (whether it be an office or a building site). Overall control is of course management's responsibility but every operative has a major part to play in their own workplace safety.

Accidents are caused by unsafe practices and conditions. Many accidents can be prevented by identifying hazards and avoiding them. This is where Risk Assessments play a major role in managing safety.

Good housekeeping is everyone's responsibility. It can make the largest single contribution to site and office safety. By keeping the workplace tidy it reduces the risk of potentially hazardous situations.

Risk assessment

Essential to the planning and running of any work is the identification of hazards that are liable to cause harm to operatives. You should aim to make sure that no one gets hurt or becomes ill through the work process.

The following information on risk assessment is taken from an advisory leaflet *5 Steps to Risk Assessment* published by the Health and Safety Executive. The leaflet contains notes on good practice which are not compulsory but which you may find helpful. It is intended to help employers and self-employed people to assess risks in the workplace.

Step 1
Look for the hazards.
  Make a list of all the hazards you find.

Look only for hazards which you could reasonably expect to result in significant harm under the conditions in your workplace. Use the following examples as a guide:

- Slipping/tripping (e.g. poorly maintained floors or stairs)
- Fire (e.g. from flammable materials)
- Chemicals (e.g. battery acid)
- Moving parts of machinery (e.g. blades)
- Work at height (e.g. from mezzanine floors)
- Ejection of material (e.g. from plastic moulding)
- Pressure systems (e.g. steam boilers)
- Vehicles (e.g. fork-lift trucks)
- Electricity (e.g. poor wiring)
- Dust (e.g. from grinding)
- Fumes (e.g. welding)
- Manual handling
- Noise
- Poor lighting
- Low temperature

Step 2

Decide who might be harmed, and how.

List groups of people who are especially at risk from the significant hazards you have identified.

There is no need to list individuals by name – just think about groups of people doing similar work or who may be affected, e.g.

- Office staff
- Maintenance personnel
- Contractors
- People sharing your workplace
- Operators
- Cleaners
- Members of the Public

Pay particular attention to the following, who may be more vulnerable:

- Staff with disabilities
- Visitors
- Inexperienced staff
- Lone workers

They may be more vulnerable.

Step 3

Evaluate the risks arising from the hazards and decide whether existing precautions are adequate or more should be done.

List existing controls here or note where the information may be found.

Have you already taken precautions against the risks from the hazards you listed? For example, have you provided:

- Adequate information, instruction or training?
- Adequate systems or procedures?

Do the precautions:

- Meet the standards set by legal requirement?
- Comply with a recognised industry standard?
- Represent good practice?
- Reduce risk as far as reasonably practicable?

If so, then the risks are adequately controlled, but you need to indicate the precautions you have in place. You may refer to procedures, manuals, company rules, etc. giving this information.

Step 4

Record your findings.

List the risks which are not adequately controlled and the action you will take where it is reasonably practicable to do more.

What more could you reasonably do for those risks which you found were not adequately controlled?

You will need to give priority to those risks which affect large numbers of people and/or could result in serious harm. Apply the principles below when taking further action, if possible in the following order:

- Remove the risk completely
- Try a less risky option
- Prevent access to the hazard (e.g. by guarding)
- Organise work to reduce exposure to the hazard
- Issue personal protective equipment
- Provide welfare facilities (e.g. washing facilities for removal of contamination and first-aid)

Step 5

Review your assessment from time to time and revise it if necessary.

Safety training

Safety training for operatives is a prudent way of ensuring that hazards within the workplace are made known and recognised before an accident claims another victim. Such training could well be in the form of a course of seminars or short presentations. These could be from a few hours to several weeks in duration with agendas covering such things as:

- understanding health and safety law
- personal conduct
- accident prevention including good housekeeping
- falling and tripping
- harmful substances
- lifting and carrying
- what to do if an accident occurs
- fire prevention/precautions

One of the best ways of keeping safety in the forefront of operatives' minds is the skilful use of advertisements posted around the site at strategic positions such as canteens, offices, washrooms, etc. The Health and Safety Executive have produced a number of these posters aimed at different occupations within the construction industry.

Safety audit

Safety audits should be carried out on all plant and machinery to ensure that they are functioning properly and that certificates are issued to certify that they meet all the legislative requirements.

Safety committees

Managers are responsible for their own area's safety policy implementation. This is linked to ensuring that safety conditions in their departments and contracts meet the requirement of all the current health and safety laws (see Fig. 4.1).

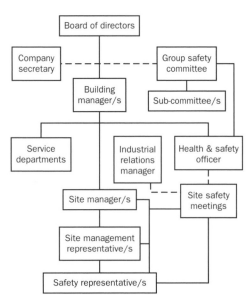

Figure 4.1 Health and safety policy organisation tree

- A Health and Safety Committee is set up to advise the directors and all employees in the maintenance of a company policy and with powers to establish sub-committees to assist the committee.
- All the work the safety committee does will be reported back to the board of company directors but in matters of urgency it may make recommendations directly to the managing director.
- The safety committee have a duty to provide technical advice on safety matters to their managers. Managers are to make all such advice known to all employees.
- Trade Unions will be consulted by the director, via the safety committee, in order to agree an implementation policy.
- It is the responsibility of the directors to ensure that their health, safety and welfare measures are constantly reviewed. It is the responsibility of managers and employees alike, to make sure that all matters relating to health and safety standards, facilities and arrangements are brought to the attention of the safety committee.

Hard hats/safety helmets

For further reading see also: the Construction (Head Protection) Regulations; the Personal Protective Equipment at Work Regulations 1992; the Health and Safety at Work etc. Act 1974; and Working Rule Agreement 21.

Without wearing a safety helmet a site operative risks a head injury. In the past injuries have ranged from slight bumps and painful bruising to irreversible brain damage or even death.

The types of injuries expected for not wearing a hard hat are:

- lacerations (many needing more than five stitches)
- fractures of the skull
- permanent brain injury (in some cases fatal)

It is now considered that all sites are to be deemed 'hard hat' areas and contractors must provide them for all operatives. Even sub-contractors must comply to a general contractor's requirement in this respect. It is the sub-contractor's responsibility to ensure that their operatives understand the safety requirements of the general contractor and that they comply with these requirements before allowing them to work on the site.

Goggles/eye protection

For further reading see: the Protection of Eyes Regulations 1974; the Protection of Eyes (Amendment) Regulations 1975; the Health and Safety at Work etc. Act 1974; and the Factories Act 1961.

The most important of our five senses is sight. We rely on our sight to confirm what our other four senses tell us. It is therefore imperative that any work operation which is liable to cause injury to the eyes has protection available supplied by the employer. Each employee/operative who is provided with eye protectors must wear or make use of such equipment. The employee/operative must take reasonable care of the equipment to ensure their safety; should the equipment become lost, damaged or destroyed then a replacement should be sought by the operative with their employer.

Operatives should be made aware of the high risks involved in not wearing eye protection when carrying out specified work processes. Eyes have enemies ranging from dust to impact or even bright lights. The commonest occupational hazards to the eyes are:

- the entry of minute dust particles
- larger gritty dust particles with sharp angular surfaces sufficient to cut the eye surface and cause severe discomfort even after removal
- high-velocity impact foreign bodies that penetrate the eye causing partial or total blindness
- chemicals entering the eye; some can destroy the cornea
- looking into bright radiations, such as welding arcs or laser beams which can damage the cornea

Work operations which require eye protection

- Cutting of wire or metal strapping under tension
- Using compressed air to remove swarf, dust, dirt or other particles
- The breaking, cutting and/or drilling of glass, hard plastic, concrete, fired clay, plaster, slag, stonework, brickwork, blockwork, bricks or tiles
- The breaking of metal by means of a hammer
- The injection of liquids or solutions by pressure
- Maintenance work involving working with substances which are dangerous to the eyes, such as acids, alkalis and corrosive substances
- Any work involving the use of molten lead
- The use of high-speed metal sawing, abrasive or disc cutting
- The chipping or scurfing of paint, slag, rust or other corrosion from metals and other hard material
- The chipping of metals
- Using hand-held cartridge-operated fixing tools
- The striking of hardened masonry nails by means of a hammer or power-driven tool
- Cleaning by means of high-pressure water jets
- Shotblasting using compressed air and abrasive materials

Ear protectors

Ear protectors should be supplied by the employer and used by construction operatives involved in work procedures that involve the making of noise greater than 90 dB(A). Hearing protection should reduce noise levels at the wearer's ear to below 90 dB(A).

Any danger to hearing depends on how loud the noise is and how long the operative is exposed to it. A general rule is that if you have to shout to be heard by someone two metres away then ear protection should be worn.

Masks, gloves and overalls

These are the three top items, according to HSE literature on the Personal Protective Equipment at Work Regulations 1992, in preventing accidents. It is the responsibility of all employers to provide such equipment to employees where their work requires it.

Safety strategies (safe person, safe place)

Safe place

Safe working conditions should be controlled by using the following observations as a form of self-assessment criteria for the workplace:

- Are working conditions appropriate?
- What protective measures are necessary (if any)?
- Disposal and frequency of waste material clearance
- What personal hygiene facilities exist?

The danger symbols most commonly used in the construction industry are shown in Fig. 4.2.

Self-assessment task

Study Fig. 4.3 and mark areas (use a soft pencil) where work operations contravene safety regulations.

Physical methods of managing safety

Before we consider physical methods of managing safety it is necessary to look at the areas and types of accidents that happen on construction and civil engineering sites.

Causes of construction accidents

Persons falling from heights

1. Falls of persons from scaffolds, including during erection, alteration, dismantling. Trestle and ladder scaffolds and scaffold collapse.
2. Falls of persons from steps and ladders.
3. Falls of persons through fragile roofs.

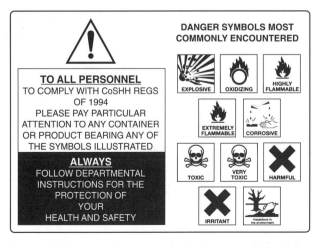

Figure 4.2 Common danger symbols (courtesy of Stocksigns Limited, Redhill, Surrey

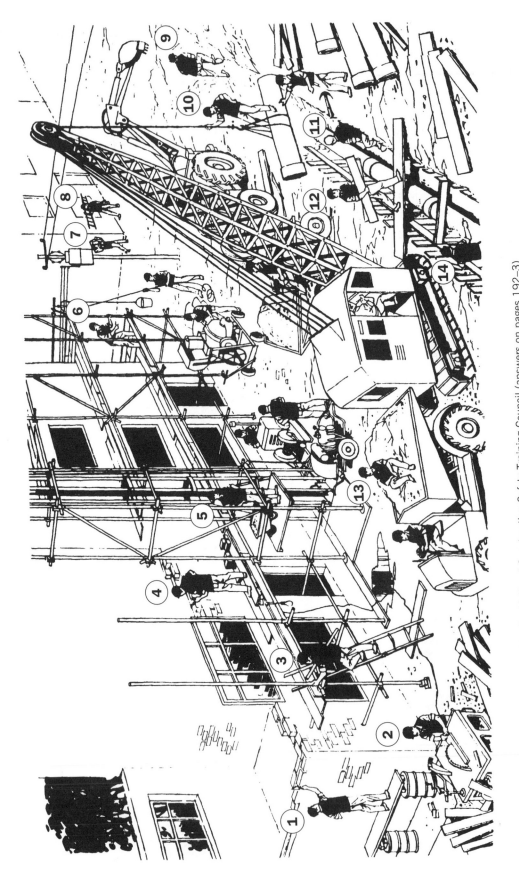

Figure 4.3 Identify the faults – designed and produced by the National Construction Safety Training Council (answers on pages 192–3)

4. Falls of persons from sloping roofs, etc.
5. Falls of persons through openings in floors, walls and down stairs.
6. Falls of persons from cradles, boatswains chairs, etc.
7. Falls of persons from heights miscellaneous, e.g. demolition, from hoists or into hoist shafts, into water, into excavations and from other working places.

Others

8. Falls of persons on the flat due to stepping on, or striking against objects, carrying or moving materials, etc.
9. Falls of materials striking persons from scaffolds, cradles, steps, ladders through fragile roofs, sloping roofs, openings, etc.
10. Excavation and tunnelling caving in.
11. Stepping on and striking against objects, protruding nails, low lintels, etc.
12. Hand tools (not power driven or cartridge).
13. Machinery, lifting equipment, hoists.
14. Transport (rail and non-rail).
15. Others; explosions, fire, etc.
16. Electricity.

Notices

It is a legal requirement to display certain notices on site. These are mainly abstracts from the Acts relevant to the type of construction work being carried out.

Prescribed notices

The following notices should be displayed, either on site or at the employer's yard, shop or offices at which employed persons attend for the work operations applicable:

- Health and Safety Law Summary which is in a poster form (Ref: ISBN 011 701424 9). This poster is required to be displayed on site, at depots and Head Office. The name and address of the enforcing authority, and address of the HSE's Employment Medical Advisory Service should be entered on it.
- The Construction (Health, Safety and Welfare) Regulations 1996, SI No 1592
- The Construction (Lifting Operations) Regulations 1961, SI No 1581
- The Construction (Head Protection) Regulations 1989, SI No 2209
- The Construction (Design and Management) Regulations 1994, SI No 3140
- Health and Safety (First-Aid) Regulations 1981, SI No 917
- The Manual Handling Operations Regulations 1992, SI No 2793
- The Personal Protective Equipment at Work Regulations 1992, SI No 2966
- The Provision and Use of Work Equipment Regulations 1992, SI No 2932
- The Management of Health and Safety at Work Regulations 1992, SI No 2051

The following notices are required to be displayed according to the type of works in progress on site:

- The Highly Flammable Liquid and Liquefied Petroleum Gases Regulations 1972, SI No 917
- Work in Compressed Air Regulations 1958, SI No 61
- Work in Compressed Air (Amendment) Regulations 1961, SI No 1307
- Diving Operations Special Regulations 1960, SI No 688
- The Control of Asbestos at Work Regulations 1987, SI No 2155
- Ionising Radiations (Sealed Sources) Regulations 1969, SI No 808
- The Reporting of Injuries, Diseases and Dangerous Occurrences Regulations 1985, SI No 2023
- The Gas Safety Regulations 1972, SI No 1178
- The Gas Safety (Installation and Use) Regulations 1994, SI No 1886
- The Control of Lead at Work Regulations 1980, SI No 1248
- The Asbestos (Licensing) Regulations 1989, SI No 1649
- The Electricity at Work Regulations 1989, SI No 635
- The Noise at Work Regulations 1989, SI No 1790
- The Lifting Plant and Equipment (Records of Test and Examination etc.) Regulations 1992, SI No 195
- The Supply of Machinery (Safety) Regulations 1992, SI No 3073
- The Health and Safety (Display Screen Equipment) Regulations 1992, SI No 2792

Examples of notices for display on site

Guidance to regulations signs

Guards and supports

Guards are provided as a physical barrier to prevent accidental touching of moving parts of machinery which are liable to cause physical injury to operatives.

A typical example is the provision of the crown guard to a circular saw bench. This is provided to prevent the wood machinist making bodily contact with the fast moving saw teeth while pushing timber past the revolving saw blade.

In the same example the wood machinist would use a specially designed support item known as a 'push stick' to prevent the fingers getting anywhere near the saw teeth when the end of the timber has to be pushed past the saw blade. Another support item would be the deadman used for the timber support on the output side of the saw bench.

Students should know that all guards provided should be fitted to all machines when in use by an operative. Guards can only be left off when the machine is isolated and not working; such as for repair, adjustment or periodic maintenance. All guards must be coloured bright yellow to signify their importance as a means of safety.

Self-assessment task

When next in a workshop note the machinery with safety guards and discuss their relevance to safety.

Good housekeeping

Unfortunately the climatic conditions on a construction site are beyond our control. In the winter months the environment to work in is usually of soft mud/slurry (especially on a site where tracked or wheeled plant churns up the clay subsoil) and in the summer months there are dry dusty conditions.

It is a well-known fact that building sites that have no policy for control of materials handling usually end up with a high amount of materials wastage. This is not only bad practice but is very costly as the materials will have to be replaced.

Typical examples are:

- Facing bricks stacked on clay soil with the lower quantities lost because of sinking in the mud. All remaining bricks are splashed with mud by passing site traffic; this will stain them and make them unusable for facings.
- Blocks stacked and damaged in a similar manner to bricks.
- Plasterboard stacked in the open with just a plastic sheet for protection from the elements.
- Timber stacked on the ground – liable to be clipped or run over by site traffic. When a carpenter requires timber the stack usually becomes unstable thus falling over and causing further obstructions with a high incidence of being damaged further.

Duty of care to the public

- Mud tracked from a construction site on to any highway is an offence. Cleaning facilities must be employed to remove it as slippery conditions will prevail in wet weather.
- Site boundaries should be maintained by repairing fencing and hoardings to prevent trespass.

Protection against falling (safety nets, cradles, harnesses)

Where the nature of the work makes it impracticable to erect scaffolding for access and egress to the work, the operatives must be prevented from falling by employing a safety harness and in some instances safety nets. These ensure that the operative has complete control of movement in order to carry out the work effectively and is restrained from falling.

Figure 4.4 Tower scaffold with outriggers for greater stability

Use of access equipment (ladders, scaffolding)

The common forms of access equipment on construction sites are:

- scaffolding (towers, independent and putlog: Figs 4.4–4.6)
- mobile elevated platforms (MEWPs)
- hoists
- ladders (Fig. 4.7)

Children on construction sites

It is not uncommon to find that children have been playing on a building site. Every reasonable step must be taken by the site contractor to ensure that children do not enter or come to harm by taking practicable reasonable steps towards their safety.

It is difficult on greenfield sites to totally isolate children from gaining access. However, areas which could

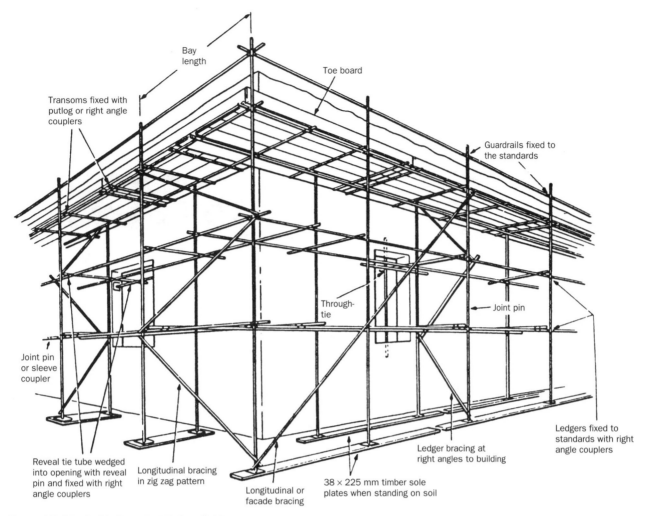

Figure 4.5 A typical independent tied scaffold

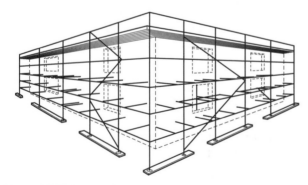

Figure 4.6 Putlog scaffold

be dangerous must be fenced off in the interest of safety. Confined sites should be made secure either with fences or hoardings to prevent trespass by members of the public.

Children are killed or maimed because to them a construction site is an inviting place to play. They have a poor awareness of danger and the site manager should always be alert for the potential danger areas which include:

- falling into holes
- being drowned
- being trapped by falling earth
- access to parked site vehicles
- injury from materials falling

It is estimated that at least twenty child deaths a year occur on construction sites with the first three items listed above being the largest single causes. This is similar to the number of adults killed in some high risk areas of construction work.

Methods of public/child protection

- erection of perimeter or localised fencing (not less than 2 metres in height)
- fencing should be of a type to prevent it being climbed
- all support posts should be securely anchored
- access should be by gates and securely padlocked when the site is unoccupied
- materials should not be stacked close to a fence providing easy access by climbing

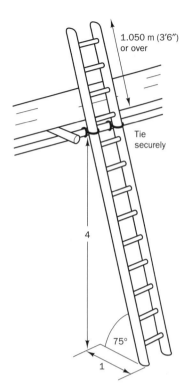

Figure 4.7 Ladder placed to reduce the risk of slipping

- a suitable warning notice should be displayed on the front of the hoarding/fencing explaining that a construction site is a place of danger for members of the public

Other areas which need special consideration and should be considered on their merits are:

- precautions where site perimeter hoarding is impracticable
- fencing off excavations or filling them in daily
- ensuring that chemicals are kept under lock and key, preferably in a compound/store
- machines and plant are safely stored in a compound or effectively immobilised when left out unattended
- material stacks to be safe to prevent easy displacement, e.g. not stacked too high
- electrical supplies not in use to be isolated at a lockable building/enclosure
- access to elevated areas should be removed at the end of the day's work, e.g. ladders removed to a storage compound

Fire precautions

A fire will only start if there is a readily available supply of:

- flammable material
- a plentiful supply of oxygen to keep the combustion process going

In fact the putting out of fires is aimed at the reverse of these by eliminating the supply of oxygen and preventing the fire spreading to other source material. It is a fact that some fires have managed to extinguish themselves only because they happened in a confined space and used up all the oxygen in the surrounding atmosphere.

Fires are now classified into different BS (British Standard) classes:

- CLASS A fires involve solid materials leaving glowing embers
- CLASS B fires involve liquids or heat liquefiable solids
- CLASS C fires involve gases
- CLASS D fires involve metals

Electrical fires no longer have a BS classification as electricity is not regarded as a fuel.

In a workshop or office the threat of a fire happening is probably the greatest potential danger of all.

Fires can be caused by:

- persons smoking
- faulty electrical equipment
- flammable liquids and vapours
- dust explosions
- spontaneous combustion

On a construction site the most common causes of fires are:

- use of unsuitable or badly positioned heating appliances (usually gas) in accommodation huts
- improper storage of highly flammable liquids and gases
- misuse or faulty LPG (liquid petroleum gas) equipment
- wet clothing which has been placed over, or fallen on to, the heating appliance intended to dry it
- carelessly discarded cigarettes and matches
- children playing with fire on site outside normal site working hours
- welding sparks from welding equipment
- improper burning of rubbish or debris
- malicious or intentional starting of fires
- faulty or wrongly used portable electrical tools/apparatus
- site vehicles
- spread of heavy vapours to an ignition source some distance away e.g. into drains or trenches, etc.

Classification and examples of flammable liquids used in construction

- **Highly flammable** is a term generally applied to liquids having a flash point below 22 °C
- **Flammable** is a term generally applied to liquids having a flash point between 22 °C and 66 °C

Examples of highly flammable liquids include:

- petrol
- cellulose solutions
- acetone
- methylated spirit
- some paint strippers
- rubber solutions

Examples of flammable liquids include:

- diesel or gas oil
- creosote

- timber preservatives
- paraffin oil
- turpentine
- white spirit or turpentine substitute

Smoking is now considered an antisocial practice and is banned in most workplaces because of other workers suffering from the effects of passive smoke inhalation.

The most likely cause of electrical fires is the incorrect fuse rating for equipment. A fuse is designed to be the weak link in an electrical circuit thus preventing a power overload should a fault occur within the equipment. It is not uncommon to find safety plugs fitted to equipment with a 13 amp fuse fitted where the safe loading should be 3 or 5 amps.

Vapours, usually spirit based, if allowed to build up within the working atmosphere, are a potential source of explosive mixtures. If the vapour mixture build-up is allowed to reach high proportions then it only takes a spark (electrical or static) to ignite the mixture to cause an explosion/fire and/or structural damage. Vapours are also harmful to the respiratory system. Any work involving the giving off of vapours should always be carried out in external or well ventilated areas. Only the minimum amount of flammable liquids should be kept, with all containers tightly closed. Where containers larger than 500 cc are stored they must be kept in a 'Highly Flammable Store' such as a suitable metal cabinet, etc.

Dust explosions are mostly associated with factories where industrial processes produce very fine particle powder dust (almost like flour grains), such as fine sanding machines with the dust being taken away at the source by extract ducting/silo equipment. Again, any spark would cause an explosion with catastrophic effect.

Spontaneous combustion is where stored materials can ignite on their own without warning. A typical example of this is where storage conditions and ambient atmospheric conditions contribute to causing an internal heat making process to establish itself, thus becoming a potential fire source. As an apprentice the author can remember picking up a very hot and damp hessian sack full of tightly packed wood chips which was at the bottom of a pile stored in an outside environment but shielded under the soffit of a staircase.

A great deal can be done to minimise the likelihood of a fire occurring but it is also important to ensure that all staff know exactly what to do in the event of a fire.

Fire procedure and means of escape

All corridors, stairs and other routes to fire exits, including the exits themselves, should be completely kept clear of obstructions.

It is good practice for employers to ensure that all staff have an induction process in what to do in the event of a fire being discovered:

- call the fire brigade
- know the exact location of the nearest alarm point
- know how to sound the alarm
- know what to do when the alarm sounds, e.g. close all windows and doors
- know where the nearest fire exits and alternative means of escape are
- know where the fire fighting equipment is and how to use it

It is essential that the Fire Brigade is called before anyone attempts to tackle a fire as valuable time may be lost in a futile attempt to put it out. The attempt may make the situation worse in helping the flames spread rapidly with complete loss of fire control.

It is important that staff should only tackle fires if it is safe to do so without endangering themselves, making sure that a quick and safe escape is always possible by staying between the fire and exit. Staff would have to know which apparatus should be selected for the specific fire they were about to tackle.

Types of fire extinguisher

- water hose reel (red in colour)
- water extinguishers (red in colour)
- foam extinguishers (cream in colour)
- dry powder (pale blue in colour)
- carbon dioxide (black in colour)
- bucket of sand (red in colour)
- BCF (short for bromochlorodifluoromethane) (olive green in colour)

In order to prevent a fire from spreading rapidly it is necessary to restrict the supply of fresh air carrying the all-important element of oxygen needed to feed the flames. In order to do this the advice is to close all windows and doors. Fire doors strategically placed throughout the building are provided to prevent the spread of smoke and heat during a fire. They also assist in reducing the oxygen getting to the flames. All fire doors will have self-closing devices (such as floor or overhead door springs) to enable them to be kept closed at all times. The practice of propping or wedging a fire door open is strictly illegal and is punishable by a fine of up to £5000. It is in the interest of your health and safety to report to the management any defective self-closing fire doors.

Water extinguishers (red)

Suitable for fires involving clothing, paper, cardboard, wood, plastic, etc.

- not suitable for live electrical fires
- not suitable for fires in close proximity to live electrical radiant heaters
- not suitable for use on flammable liquids

Foam extinguishers (cream)

Suitable for fires involving flammable liquids such as petrol, diesel, cooking oil, paint, etc.

- Not suitable for live electrical fires

Dry powder (blue), BCF (olive green) and carbon dioxide extinguishers (black)

Suitable for fires involving flammable liquids such as petrol, diesel, cooking oil, paint, and those involving live electrical apparatus.

The Fire Precautions Act 1971

Offices were brought under the Fire Precautions Act in 1977. Under the Act (and the relevant regulations) all premises where more than 20 persons are employed at any one time, or where more than ten persons are employed elsewhere than on the ground floor, require a fire certificate. The fire certificate, which must be kept in the building, specifies certain conditions relating to fire precautions. Offices which do not require a certificate are required by regulations to take certain precautions in relation to means of escape and fire fighting equipment.

Workplaces

Greenfield site

Greenfield sites are areas, as the name suggests, in open countryside. They are often referred to as rural sites. Some may travel across the countryside for many kilometres such as motorways; others may only occupy a localised large area. The scale of construction (usually to strict time schedules) can involve large volumes of soil to be moved, and large areas to be cleared in forming embankments, cuttings, bridges and tunnels. Planning of such projects requires real professional accuracy as huge machines capable of moving vast quantities mean large financial loss to the contractor for any inaccuracies in the setting out of the works. Large construction projects are usually carried out by civil engineering companies building such structures as:

- dams
- airports
- harbours
- power stations
- roads and motorways
- railways
- canals

- bridges
- sewage works
- drainage schemes
- oil refineries
- storage facilities such as oil tanks and gas holders, etc.

Another type of greenfield site could be a housing development involving a speculative developer/builder. All new infrastructure, consisting of estate roads and services such as main foul and surface water sewers will have to be put in before construction of the houses can take place.

A civil engineering sub-contractor would construct the roads and main sewers while the speculative developer/builder would build the houses and connect the foul and surface water to the main sewers previously put in by the civil engineering sub-contractor. All the other infrastructure services of gas, water, electricity, telephones, etc., will eventually be put in by the respective service authorities, crossing under the roads (in ducts) and along the footpath routes to eventually connect up within the houses via the ground floor ducts.

Confined site

A confined site is one that may well be referred to as an urban site and could cover as little as an area of approximately 200 square metres. They are usually in built-up areas where access and egress is severely limited because of the surrounding site boundaries. A hoarding will be required around any exposed site boundary in order to prevent members of the public from straying on to the site. The site has a relatively small area to be cleared, and a small amount of topsoil to be excavated and stored on site or sold and carted off site, because space is at a premium. Machines suitable for working in small locations such as a backactor/loader will be adequate for this work.

Storage of construction materials is again severely limited and it may be necessary to ask for reduced deliveries, thus increasing costs.

4.2 Measurement and costing of material quantities

Topics covered in this section are:

- Calculating material quantities from drawings
- Ordering material quantities in appropriate units
- Identifying sources of information for pricing materials
- Pricing materials to an appropriate degree of accuracy

To carry out measurements, find quantities and obtain costings the student will have to use arithmetic techniques of addition, subtraction, multiplication, division, percentages, ratios and proportions.

To carry out calculations for areas, the student will use the geometric techniques for rectangle, triangle, circle and trapezium.

To carry out calculations for volumes, the student will use the geometric techniques for cube, pyramid, cylinder and frustum.

Note that section 4.3 contains the formulas for arithmetic and geometric techniques.

Techniques used to calculate material and resource requirements

The use of linear, square and cubic dimensions are common in the construction industry. The units are the SI metric system. In the construction industry, the use of centimetres as a unit of linear measurement is not preferred and should not be used in any of the calculations that are produced. The units that are acceptable for linear measurement are metres and millimetres only.

The formulas used in this section use the following abbreviations:

A = area
V = volume
l = length of side
h = height
r = radius
C = circumference
d = diameter
a = side a
b = side b

To calculate material requirements the following formulas will be required:

Areas

Square In geometry, a quadrilateral (four-sided plane figure) with all sides equal and each angle a right-angle. Its diagonals also bisect each other at right-angles.

The area A of a square is the length l of one side multiplied by itself; therefore area $A = l^2$.

Similarly, any quantity multiplied by itself produces a square, represented by an index (power) of 2; for example $3^2 = 9$ and $8.3^2 = 68.89$.

Triangle In geometry, a three-sided plane figure. A scalene triangle has no sides of equal length. An isosceles triangle has two equal sides (and two equal angles). An equilateral triangle has three equal sides (and three equal angles of $60°$). A right-angled triangle has one angle of $90°$.

If the length of one side of a triangle is l and the perpendicular distance from that side to the opposite corner is h (the height or altitude of the triangle), its area $A = \frac{1}{2}l \times h$ (Fig. 4.8).

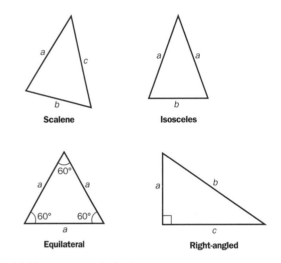

Figure 4.8 Various types of triangle

Circle In geometry, a path followed by a point which moves so as to keep a constant distance, the *radius*, from a fixed point, the *centre*.

The longest distance in a straight line from one side of a circle to the other, passing through the centre, is called the *diameter*. It is twice the radius.

The ratio of the distance all the way round the circle (the *circumference*) to the diameter is an irrational number called π (*pi*) roughly equal to 3.14159. A circle of radius r and diameter d has a circumference C equal to πd or $C = 2\pi r$, and an area $A = \pi r^2$ (Fig. 4.9).

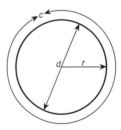

Figure 4.9 A circle

Trapezium In geometry, a four-sided plane figure (quadrilateral) with only two opposite sides parallel.

If the parallel sides have lengths a and b respectively and the perpendicular distance between them is h (the height of the trapezium), its area $A = \frac{1}{2}(a + b) \times h$ (Fig. 4.10).

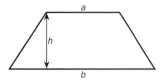

Figure 4.10 A trapezium

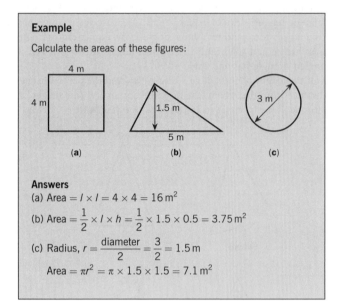

Example

Calculate the areas of these figures:

(a) 4 m × 4 m

(b) 1.5 m, 5 m

(c) 3 m

Answers

(a) Area $= l \times l = 4 \times 4 = 16\,m^2$

(b) Area $= \frac{1}{2} \times l \times h = \frac{1}{2} \times 1.5 \times 0.5 = 3.75\,m^2$

(c) Radius, $r = \frac{diameter}{2} = \frac{3}{2} = 1.5\,m$

Area $= \pi r^2 = \pi \times 1.5 \times 1.5 = 7.1\,m^2$

Volumes

Cube In geometry, a solid figure whose faces are all squares. It has six equal-area faces and twelve equal-length edges.

If the length of one edge is l, the volume of the cube $V = l^3$ and its surface area $A = 6l^2$ (Fig. 4.11).

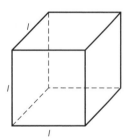

Figure 4.11 A cube

Pyramid In geometry, a solid figure with triangular side-faces meeting at a common vertex (point) and with a polygon as its base.

The volume of a pyramid, no matter how many faces it has, is equal to the area of the base multiplied by one-third of the perpendicular height (Fig. 4.12).

Cylinder In geometry, commonly interpreted as a surface generated by a set of lines which are parallel to a fixed line and passing through a plane curve not in the plane of the fixed line; a tubular solid figure with a circular base, ordinarily understood to be a right cylinder, that is, having its curved surface at right angles to the base (Fig. 4.13).

Volume = base area (circle) × length

Frustum In geometry a shape resulting when a 'slice' is taken out of a solid figure by a pair of parallel planes. A conical frustum, for example, resembles a cone with the top cut off.

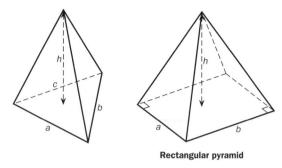

Figure 4.12 Various types of pyramid

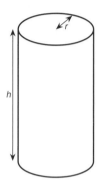

Figure 4.13 A cylinder

The volume and area of a frustum (the lower portion containing the base) are calculated by subtracting the volume or area of the 'missing' piece from those of the whole figure (Fig. 4.14).

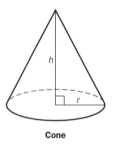

Cone

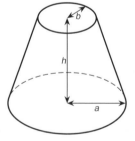

Figure 4.14 A frustrum

Geometric terms

- **Acute angle** – an angle which is less than 90°
- **Right angle** – an angle of 90°
- **Obtuse angle** – an angle that is more than 90° and less than 180°
- **Reflex angle** – an angle which is larger than 180°
- **Apex** – the highest point of a triangle

Measuring construction materials from drawings

Photocopied drawings

Beware of some photocopied drawings as modern photocopying machines have the ability to either enlarge or reduce drawings thus altering the original drawn scale size (even though the original drawing information box may state the original drawing scale). This is a problem either way as to measure distances using a scale rule on a drawing which has been slightly enlarged from the original means you will be over measuring and conversely a slightly reduced drawing will be under measuring to the stated scale.

Reading of construction drawings

Students entering the construction industry for the first time will probably find that reading a construction drawing or layout will be a daunting task and will not be sure what they are expected to be looking at. Don't be put off by this feeling as the more drawing details you see the easier it becomes to read them.

Self-assessment tasks

1. Obtain an architect's detailed drawing of a single storey structure and carefully look at what is being shown. Look for, and read the information box on the bottom right-hand corner. This will tell you about the development and will state such items as drawing number, revisions, title of drawing, who produced the drawing and when, the client and the architect.
2. Look at the drawing details to see how the structure is drawn; find sections and locate where they are taken on plan and in what direction they are looking. Note the information that is shown in section but is not possible to show on plan.
3. Now look at elevations of the structure and decide which is north, south, east and west. Try to match the positions of the windows and doors to the plans.

Now we will consider where we can get sizes for each respective element we need to obtain the quantities for. The drawings need to be a direct copy of the originals and should contain **plans**, **elevations** and **sections** including a **block** or **site plan** showing the area of the proposed development. Check the scale against the figured dimensions shown on it.

Beware that the drawing's scale may differ, with more than one scale being used on the drawing. For example, plan scales may be smaller than the scales used to show sections. All scales should be stated on the drawing, so look for this information first and study the drawings to decide which details refer to which scale.

With this information it should be possible to prepare detailed estimates of the quantities of materials required for the structure.

Before we can do this there are golden rules the student must always bear in mind. These are linked to the points made under 'photocopied drawings'.

1. Where figured dimensions are shown on a drawing they should always be used in preference to scaling. So always look for figured dimensions first.
2. Only use scale dimensions when figured dimensions are missing and only then when you are sure that the scale on the drawing is correct. To test the correctness of scale you should attempt to prove the drawing by using a scale rule

measuring a known dimension on the drawing (this could be anything that you know the finished size of).

If you are ever in the position to make photocopied drawing details to give to another then do make sure the recipient can determine scaled dimensions with a degree of accuracy. One method of ensuring this is to draw the scale on the original drawing; this then increases or decreases in proportion to the details being photocopied. It does mean of course that the original stated scale in the information box may be incorrect but at least the recipient can work with some accuracy. (Figure 4.25 shows an example of this.)

Material requirements calculated from drawings using geometric techniques

Materials in the construction industry are commonly ordered in the following units:

- by **volume** as 'cubic metres' = m^3
- by **length** as 'linear metres' = m
- by **each type** as 'each' = ea.
- by **quantity** as 'number' = No.
- by **sheets** as 'number' = No. of sheets followed by their size
- by **weight** as 'units' = e.g. tonne, kilogram or gram
- by **liquid measure** as 'litres' = ltrs.

The calculations for materials measured in square metres often require adjustment because of openings. A prime example of this would be when measuring the amount of brickwork for an elevation. The overall length (from plan) and overall height (from section or elevation) can be measured and if we square out the dimensions (the gross area) we are actually allowing for brickwork over the window and door openings; this therefore is in excess of requirements. To correct the quantity it is necessary to deduct out the areas occupied by the windows and door/s (giving net area).

The rules are therefore:

- measure and calculate the gross area
- make adjustments to allow for openings to give the net area (actual area of material required)

Example

Calculate the area of brick needed for the wall shown in Fig. 4.15.

Answer

To calculate the area of material required, we need to take account of the area of the openings, i.e. the door and windows. First calculate the gross area of the wall (Fig. 4.15a):

$$\text{Area} = \text{width} \times \text{height} = 5 \times 2.5 = 12.5\,m^2$$

Then calculate the area of the door and windows:

$$\text{Area of door} = 1 \times 2 = 2\,m^2$$

$$\text{Area of window} = 1.5 \times 1 = 1.5\,m^2$$

$$\text{Total area of openings} = 2 + 1.5 + 1.5 = 5\,m^2$$

Finally, subtract the area of the openings from the gross area to give the area required (Fig. 4.15b):

$$\text{Area required} = 12.5 - 5 = 7.5\,m^2$$

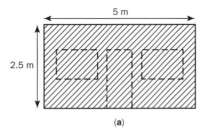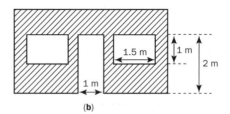

Figure 4.15 Calculation of area of walling and deducting out openings

This method is classed as measuring in gross and deducting out to give the net area. A quantity surveyor uses this system as errors are minimal because the maximum quantity of material is measured first thus preventing shortages of ordered materials. Over measure is only possible if insufficient deducting out has occurred.

Material data

Self-assessment tasks

Using your research skills:

1. Find out as much information as possible on the following material requirements, e.g. thickness, size and length of:
   Facing brick
   Lightweight blocks
   Plasterboard
   Copper pipe
   Timber
2. Find out as much information as possible on the following material requirements, e.g. volume, sizes and weights of:
   Cement
   Sand
   Aggregate

Calculations for waste

Waste is inevitable when materials are handled, transported, offloaded, placed and cut. The amount of waste caused by damage will depend on the mode of handling/transportation used. Waste can be referred to as:

- storage – waste in stacking, handling, transportation to craftsperson
- residual – waste left over after fixing as unused material
- cutting – waste left over after cutting up materials

Handling of any building material will increase the chances of damage each time it is picked up or moved. It is therefore important that materials are delivered in pristine condition, with unloading and storage as close to the work operation as possible with minimal handling required.

Waste is measured as a percentage with typical wastage factors ranging from 2% to 25%. The waste factor will vary between different materials. The softer the material a product is made from the easier it is to be physically damaged. For instance the brick examples shown below have wastage percentages ranging from only 2% for the very hard bricks to 5% for the less dense brick. Hollow blocks on the other hand, being less dense with cavities and thin walls, are more susceptible to damage – wastage factor would be up to 10%.

Example

Common bricks cost £96 per thousand. What would be the total cost per thousand bricks for a job, assuming 5% are lost through wastage and damage?

Answer
First find 5% of £96. To do this divide £96 by 100 to give 1%, and then multiply by 5:

$$5\% \text{ of } £96 = \left(\frac{96}{100}\right) \times 5 = £4.80$$

The total cost allowed for 1000 bricks is given by the cost for 1000 plus the wastage cost:

$$\text{Total cost} = £96 + £4.80 = £100.80$$

Costs for engineering bricks and facing bricks, including wastage, are given below. Using the method above, you should be able to check that the values given are correct.

- Engineering bricks Class A @ £470/1000, 2% waste = £479.40/1000
- Engineering bricks Class B @ 180/1000, 2% waste = £183.60/1000
- Facing bricks @ £200/1000, 5% waste = £210.00/1000

With timber the cutting waste can be reduced considerably by ordering a single length to cut two or more pieces from it. For example the lengths 2.9 m and 1.5 m would give an overall length of 4.4 m with only 100 mm waste when ordered from a 4.5 m length. If the lengths were ordered individually the lengths ordered would be 3.0 m and 1.8 m respectively. The waste would be as two offcuts of 100 mm and 300 mm, making a wasted length of 400 mm.

Examples of finding true costs

True costs of each material can be found by:

- knowing the current price paid for the material (which should include cost of delivery to site)
- adding a percentage for damage when unloading
- adding the cost of labour to unload (time taken × labourer hourly rate)

Note that prices quoted in this section may not be current prices, therefore the student should check current information sources if assignments are set requiring costings.

Portland cement in bags @ £68.00/tonne, allow 5% waste, labour cost of unloading £3.92/tonne; the total cost to the contractor is £75.32/tonne to unload and store on site.

Hydrated lime in bags @ £83.00/tonne, allow 5% waste, labour cost of unloading @ £3.92/tonne, total cost = £91.07/tonne.

Building sand to BS 1200 @ £7.08/tonne, allow 10% waste, no cost of unloading necessary as ordered on a tipper lorry, total cost = £7.79/tonne.

Self-assessment task

Research current hourly rates for the following trades:

- Labourer
- Bricklayer
- Carpenter
- Plasterer
- Plumber
- Electrician

Self-assessment tasks

1. Which of the following softwood timber sizes are non-standard timber sections?
   100×22 mm; 115×45 mm; 200×22 mm; 175×100 mm.
2. If you were requested to order the following timber lengths, what would be the lengths you would actually order from the timber merchant?
   1.850 m; 3.333 m; 6.900 m; 2.225 m.

Sources of information for pricing materials

The sources for obtaining prices can come from three main areas:

- manufacturers' price lists
- builders merchants
- standard pricing books/lists

The following well-known publications should be sought for reference:

- Wessex Database Building (also on a computer based system)
- Spon's Architects' and Builders' Price Book
- Hutchins' Priced Schedules

Calculations using timber

The basic lengths of sawn softwood are:

1.8 m 2.1 m 2.4 m 2.7 m 3.0 m 3.3 m 3.6 m 3.9 m
4.2 m 4.5 m 4.8 m 5.1 m 5.4 m 5.7 m 6.0 m 6.3 m

Softwood timber lengths can easily be remembered as they increase each time by 0.3 m (300 mm) increments. Note the minimum length (1.8 m) and maximum length (6.3 m).

Table 4.2 Sawn softwood timber: basic sizes as supplied by timber merchants

Thickness (mm)	Width (mm)									
	75	100	115	125	150	175	200	225	250	300
16	•	•		•	•					
19	•	•		•	•					
22	•	•		•	•					
25	•	•		•	•	•	•	•	•	•
32	•	•	•	•	•	•	•	•	•	•
40	•	•	•	•	•	•	•	•	•	•
45	•	•	•	•	•	•	•	•	•	•
50	•	•	•	•	•	•	•	•	•	•
63		•		•	•	•	•	•		
75		•		•	•	•	•	•	•	•
100		•			•		•		•	•
150					•		•			•
200							•			
250							•			
300										•

Table 4.3 A typical example of a price book entry for supply and fixing of timber floor joist sections

Sawn softwood (untreated)	Labour hours	Net labour price (£)	Net material price (£)	Net unit price (£)	Unit
50×100 mm	0.12	0.59	0.93	1.52	m
50×150 mm	0.13	0.64	1.41	2.05	m
50×200 mm	0.14	0.69	2.68	3.37	m
75×150 mm	0.15	0.74	3.03	3.77	m
75×200 mm	0.20	0.98	4.65	5.63	m
75×250 mm	0.25	1.23	5.29	6.52	m
75×300 mm	0.30	1.47	5.81	7.28	m
100×200 mm	0.27	1.32	5.79	7.11	m
100×250 mm	0.30	1.47	7.72	9.19	m
100×300 mm	0.36	1.76	9.54	11.30	m

Additional costs to consider

If the timber lengths are required to be supplied in one length and they fall into the following long length categories, then you will have to add the appropriate percentage on costs:

- lengths from 5.1 to 6.0 m add 8%
- lengths from 6.3 to 7.2 m add 15%
- lengths from 7.5 m and over add 20%

Other treatments are costed as follows:

- preservative treatment +15%
- stress grading GS +7%
- stress grading SS +25%
- gauging (regularising) +6%
- special sawing +15%

Pricing materials to appropriate degree of accuracy

The degree of accuracy that students are expected to price materials to is as follows:

- all measurements in millimetres (mm) are expressed in metres (m) to three decimal places
- all calculations are worked to two decimal places
- all final answers are rounded up to the nearest whole number

4.3 Planning construction operations for a low-rise domestic building

Topics covered in this section are:

- Identifying stages and sequences of construction operations for a low-rise domestic building
- Preparing a plan of construction using bar charts
- Calculating from published data the duration of construction operations
- Planning a site layout
- Planning setting out using trigonometric and algebraic techniques

Stages and sequences of construction

On a large construction project where a number of low-rise dwellings are to be built, the stages of construction would be:

1. Main sewers (foul and surface water)
2. Roads
3. Sub-structure
4. Superstructure
5. Site handover to Local Authority (adoption of roads and main sewers for future maintenance)

The only two stages (Fig. 4.16) we will consider in the process of constructing a low-rise domestic building are:

- sub-structure
- superstructure

It is assumed that the main sewers and road already exist.

Sub-structure is the area involving the foundations, walls up to dpc level, house drainage and underground services.

Superstructure is the structure above the dpc course.

Construction projects at first may seem a daunting task when you consider that their success depends upon the skilful employment of the following resources:

- finance
- labour
- machines
- time
- materials

and to this end professional occupations exist to organise, plan, control, inspect and produce a suitable structure for the client. How they employ finance, labour, machines, time and materials depends upon the nature of the construction process. The nature of the construction process is best described as:

- traditional construction or
- specialist construction

Whatever the type of construction process, the way the structure can be built will depend on the sequence in which the elements can be constructed. Craftspersons are required to attend the site and construct their particular element according to their trade. To do this successfully these trades will need to know when they will be required on site to carry out their work. To achieve this, we need to consider each particular trade as a separate set sequence in the construction process. These sequences are known as:

- sequential operations
- non-sequential operations

It is impossible to support the roof without building the walls first, and it is an equally impossible idea to support the walls without first constructing the foundations. Therefore, the structure's sequence of operations is set mainly to a precise order known as 'sequential operations'.

There are some work operations which are not dependent on this set order and they can be carried out totally independently of any set sequence. These are the non-sequential operations.

For this unit we shall only consider the low-rise domestic building which is of a traditional construction, consisting of trench fill (concrete) foundations, brick/blockwork external cavity walls, concrete ground floor, and a pitched roof using prefabricated roof trusses.

In planning any project, the student needs to consider the requirements and to put them in some chronological order to suit the construction process. By producing a list of all the trade operations in a chronological sequence the student will be entering the planning process by providing basic planning information. This is the beginning of providing an overall programme of works for the whole of a project by:

- mapping out a chronological sequence of trade operations
- highlighting and understanding the interrelation of all trades, which is an essential part to any planning process

Sequence of construction operations for a low-rise dwelling

(Traditional construction with a concrete ground floor. Estimated time can be filled in as part of the self-assessment task on p. 174.)	Estimated time in days?

1. Contractor takes possession of site
2. Set up – site huts, compound, fences/hoardings, support racks for materials, etc.

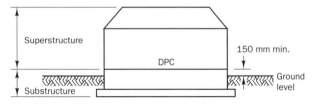

Figure 4.16 Definition of sub-structure and superstructure

	Estimated time in days?

3. Strip site – topsoil to a depth of 150 mm minimum
4. Excavate to reduced level
5. Set out house foundations
6. Excavate foundation trenches (including ducts for services)
7. Cast concrete to deep strip foundations
8. Brickwork to dpc (damp-proof course) level
9. Fill cavity below dpc level
10. Supply and compact hardcore to ground floor slab (including 50 mm sand blinding)
11. Lay dpm (damp-proof membrane)
12. Spread and level concrete to ground floor slab
13. Excavate trenches for drainage/inspection chambers
14. Lay drainage pipes in trenches and construct inspection chambers
15. Test drains for leaks and backfill drainage trenches
16. Block out with scaffold at dpc level
17. Bed dpc in mortar
18. Start brickwork to first lift (finishing approx. 1.5 m above dpc level – 21 courses)
19. Scaffold to first lift (including preparing safe support to foot of scaffold)
20. Start brickwork up to first floor joist level
21. Fix in position first floor joists
22. Complete brickwork above floor joists (end of second brickwork lift)
23. Scaffold to second lift
24. Start brickwork to third lift
25. Scaffold to third lift
26. Erect roof (including fixing fascias and soffits)
27. Scaffold to fourth lift
28. Brickwork topping out (including chimney construction)
29. Felt, batten and tile roof
30. Fix lead flashings (plus full decoration of fascia, soffit and gable verge)
31. Fix external guttering and rwp (rainwater pipe)
32. Strike scaffold
33. Plumber first fix (positioning pipe runs)
34. Carpenter first fix (stud partitioning, flooring, ceiling noggings)
35. Electrician first fix
36. Tracking (fixing of plasterboard to ground and first floor ceilings)
37. Plastering
38. Floor screed
39. Glazier
40. Wall tiler (kitchen, bathroom, shower, etc.)
41. Plumber second fix
42. Carpenter second fix
43. Electrician second fix
44. Painting internal
45. Painting external
46. Carpenter final fix (door furniture, etc.)
47. Floor tiling (complete ground floor)

	Estimated time in days?

48. Cleaning operations
49. Snagging (inspection and listing of all trade defects)
50. Trade snagging operations
51. Concrete paths and driveway
52. Fencing to boundaries
53. Landscaping (spread and level topsoil, grass seeding, planting, etc.)

Explanatory notes on sequence of construction

Items 1 & 2 These are items that are referred to as preliminary works as they are non-productive but essential to the site's development.

Items 3 & 4 Excavation of topsoil and the additional depths to the reduced level excavation are required to extend at least 1 m past the overall size of the structure only. Stripping the complete site is uneconomic and will not be required.

Item 5 The setting out is relative to the site layout as the correct position of the property has to be set out as agreed with the Local Authority.

Items 6 & 7 Access ducts for the incoming services need to be put in and passed through the foundation trenches before the foundations are cast. In firm subsoil the sides of excavations for the foundations act as a means of vertical support to the wet concrete for anything up to four hours, allowing it time to set. The setting out of foundations must be accurate as all walls must be positioned centrally on them to give correct structural support.

Items 8, 9, 17, 18, 20, 22, 24 & 28 These refer to a typical gang of five bricklayers to two labourers.

Items 10, 11 & 12 Before a concrete ground floor slab can be cast, it is necessary to provide a firm base for its support. Therefore hardcore has to be compacted and finished off with a 50 mm sand blinding. The sand blinding is required to provide a smooth surface over the sharp angular hardcore ready to receive the damp-proof membrane (item 11).

Items 13, 14 & 15 Drainage is best put in before the structure is completed as access for mechanical plant (excavator and dumper) is easier without the walls providing obstruction. It also helps in the taking and transferring of drainage levels as an unobstructed view is essential when using a surveyor's level.

Item 16 This is necessary to provide a firm level standing for the bricklayers to start the building of the external walls above the dpc.

Items 19, 23, 25 & 27 Lifts of brickwork are usually 21 courses (1.5 m). This is considered to be the maximum height that a bricklayer can comfortably work without over reaching. All lintels, windows and door frames are built into the brickwork as the work proceeds. A lift of scaffold refers to the need to raise the scaffolding platform to suit the growth of bricklayer's work.

Item 21 The carpenter is required at this point to lay out and position the first floor joists at 400 mm centres on top of the internal skin of the external cavity walling. Once this is done the bricklayers then fill in the gaps on each side of the joists with the blockwork.

Item 26 The carpenter has to fix the roof trusses in their required positions spaced at 600 mm centres. To support the trusses the bricklayer would have bedded in mortar a timber wallplate on top of the finished internal skin of blockwork. Once the roof trusses are fixed, the carpenter then has to form the eaves/verge fascias and soffits.

Item 29 At this stage of the construction process the property is considered to be watertight and the internal structure is protected from the external environment. Internal trades can now begin.

Item 30 With the roof finally on the structure the plumber can now complete the final stage of making the structure watertight. This is done by fixing lead flashings to the chimney stacks and a lead sleeve to the soil and ventilation pipe projecting through the roof tiles.

Item 31 The plumber is also involved in dealing with getting rid of the roof surface water by installing the rainwater guttering and downpipes.

Item 32 'Striking' of a scaffold is the term used for taking it down on the completion of works requiring scaffold access.

Item 33 Although the heating system is not ready for fixing at this stage, it is necessary for the plumber to fix in position the pipe runs for the heating and cold water supply which will eventually be covered up, e.g. in floors and walls.

Item 34 First fix carpenter is an important stage internally as once the staircase and first floor boarding is fixed the rooms begin to take shape on the first floor using timber stud partitioning and the fixing of ceiling noggings ready for fixing the plasterboard ceilings.

Item 35 Cable runs, conduits and fixing boxes need to be put in before they are covered up by the plasterboard tacking.

Item 36 Ceilings and partitions are faced-up with dry plasterboard sheets. Fixing is by rust-proofed nails, usually galvanised.

Item 37 Plastering is classed as a wet trade. The water used in the mixing of the plaster mixture will take some weeks to dry out. This drying out time will vary according to the time of year. Usually on average this period is about 3 weeks; you could expect it to be longer in the winter and shorter in the summer months.

Item 38 Up to now the ground floor finish has been left as a tamped surface of the concrete slab. The finished ground floor level is 50 mm above the slab level and the reason why the screed has not been cast is because it would have been severely damaged by all the construction operations that have already taken place. A floor screed is a semi-dry cement/sharp sand mix having very little water added.

Item 39 All glazing should be left as long as possible because of the threat/temptation of vandalism. If it is found necessary to prevent the elements from entering the structure then temporary heavy duty polythene provides the answer and is a fraction of the cost of replacing glass.

Item 40 Wall tilers are a trade that specialise in the application of ceramic glazed wall tiles to kitchens, bathrooms

and showers, etc. Some builders insist the walls are tiled before the plumber second fixes, while others prefer walls tiled after the plumber has second fixed.

Item 41 Second fixing for the plumber consists of fixing and connecting up WCs, baths, basins, sink, taps, waste connections, etc. by picking up connections to pipe runs already put in during plumber first fix.

Item 42 The carpenter's second fix operations consist of hanging doors, fixing kitchen units, architraves, skirtings, linings, etc.

Item 43 The electrician's second fix operation consists of fixing all electrical apparatus such as light fittings, switches, power points, etc. to cables and boxes previously put in during the first fix operation.

Item 44 Internal painting is the last major trade to take place within the structure. Obviously the property should be dry enough for painting. The time of the year will dictate this. It is considered to be good practice in newly constructed properties that all internal plastered walls are emulsion painted and not papered. Emulsion paint allows the wall surface to breathe. Papering walls in a new property prevents the walls from breathing out the water absorbed through the application of the wet trades and is liable to cause a severe case of mildew/mould to appear internally. It takes at least 12 to 18 months for the inside of a property to gradually dry out after construction

Over this period of drying out it will be observed and be a source of complaint by the occupant that cracks appear around the internal surfaces of the structure. This is quite normal shrinkage cracking and is often wrongly suspected by the tenant to be of a structural nature and a cause for concern.

Item 45 External painting can be carried out at any time, weather permitting (usually from April through to September) and providing other trades are not around to damage the finished surfaces.

Item 46 The carpenter has to final fix the items left off, such as door knobs/handles, push plates, escutcheons, etc. to allow the painters access to complete surfaces.

Item 47 A floor tiler is required for the laying of the ground floor tiles throughout.

Item 48 As each trade is responsible for their own rubbish to be cleared as their works proceed there should be little or no cleaning in theory to be finalised at the end of the works. This is not actually so as a final clean with polishing is necessary to give the appearance of a finished product, and a specialist firm of cleaners is employed just to give the finishing touches.

Item 49 Snagging is the final process of ensuring that each trade has left the property in a fully functioning order. It is usually carried out by the site foreman inspecting every craftsperson's work to ensure that it complies with the standards of workmanship required. A list of snagging items (faults) is produced for each trade to put right before the property is handed over to the client.

Item 50 Time is given to each trade to put any snagged items right.

Item 53 Landscaping consists of spreading and levelling the topsoil previously stored on site following the Item 3 operations. This will also include the sowing of grass seed or alternatively the laying of grass turfs and planting of shrubs, etc.

Self-assessment tasks (see page 170)

1. Look at the sequence of construction item numbers (1–53) and list those that you consider to be sequential.
2. Those that are left must be non-sequential – now look to see where the non-sequential operations could be slotted in for their earliest work start.
3. List the items that make up
   a) the sub-structure
   b) the superstructure

Calculating the duration of construction from published data

Useful sources are to be found in a library/learning centre containing information on trade rates, material costs, machine outputs, etc.

The following well-known publications should be sought for reference:

- Wessex Database Building (also on a computer-based system)
- Spon's Architects' and Builders' Price Book
- Hutchins' Priced Schedules

If you can calculate the total quantity of materials and you know the work rate it is possible to calculate the duration of that work element. Coupled to this, if you know the current craftsperson's hourly wage you can cost the whole works. Some work time elements may not be known and a pure 'guestimate' will have to be taken by the quantity surveyor at the time of estimating for the building works. On the other hand, if the builder has carried out similar work tasks on other jobs then the figures ascertained from the past work will assist in formulating estimating data/work time elements allowing more accurate costing which is more reliable than a 'guestimate'.

A contractor may employ the services of a work study engineer to ascertain the true cost of the works by actually observing the work task in progress on site.

Site layout is planned

Before any programme of works is produced, certain aspects relating to the site have to be considered. The intended construction process will play a major role in the organisation of the site before and after work gets underway. It is therefore important that the site layout is pre-planned to take into consideration the following:

- work areas
- site huts
- security
- storage of materials (site sloping or level)
- adjacent buildings
- site access
- site egress
- turning circles
- types of plant
- existing services
- site temporary services
- hardstandings for material deliveries

Site huts Used for offices/storage/security for:

- clerk of works
- contracts manager
- agent
- quantity surveyor
- messing facilities
- stores
- storage containers
- compound storage and security

Services Required for connection on completion of works:

- water
- electricity
- gas
- sewerage
- telephone

Plant Required to assist the construction process:

- forklift
- digger
- dumper
- cement mixer
- scaffold

Access Required for vehicles coming on site:

- turning circle
- reversing area
- large lorries
- road access
- site routes

Egress Required for vehicles going off site to surrounding roads:

- mud tracked on to highway
- wheel washer
- labourer for road cleaning operations

Environmental Factors beyond our control:

- winter, freezing conditions
- summer, dry and dust
- rain, soft mud

Storage of materials Required for the construction process:

- bricks
- blocks
- pipes
- cement
- aggregates
- sand
- timber, joists
- timber, second fix items
- window frames
- doors and doorframes
- lintels
- roof trusses
- loose, bagged and boxed items

Ordering and delivery of materials requirements

Ask the question for each material to be ordered: 'When will this material be required on site?'.

The supply of materials is one of the most crucial factors affecting the progress of any construction operation.

However, to know when the material is required on site is a major part of the planning process which will be explained later. For example, it would not be prudent to order roof tiles for delivery during the foundation/substructure stage as storage could be a problem, together with the strong possibility that they could be stolen or damaged before they are required for fixing. The longer materials remain on site unfixed the greater the hidden cost of having to reorder.

The ideal solution is for the 'company buyer' (person responsible for buying and negotiating special rates for the materials) to place an order as early as possible with the instructions to the supplier that the materials will be called forward from the site as and when required. This way the materials are guaranteed for when the site requires them. It also obviates problems where manufacturing periods will be required for some specialist materials, which can be anything up to 12 weeks for delivery.

By now the student should be building up a mental picture that planning is one of the most important tasks of any contractor and that the site layout is not the only factor in the planning process.

Site should request the manufacturer/supplier to deliver one week before the materials will actually be used. This will prevent overcrowding and storage problems to allow a smooth-running site.

Remember, delivery periods can affect an overall programme of construction works. Many of the materials required are readily available from a stockist (such as a builders' merchant) but other specialist items may have to be manufactured by a company which is committed to production schedules for other clients. Therefore your order will be processed strictly on a chronological basis. Remember also, some manufacturing companies have a complete holiday shutdown period/s during the year which can range from anything from one to two weeks; however, the company buyer should be aware of this. During this period these companies carry out necessary maintenance work to plant and equipment which is essential for the continued manufacturing of their products.

Self-assessment tasks

1. Visit a builders' merchant and ascertain which materials are readily available from stock. Pick up a current price list of materials.
2. Make enquiries about which materials are affected by manufacturers' shutdowns and what types of materials require long manufacturing periods (together with an estimate of time).

Planning construction operations in the form of a bar chart

To consider the limitations of material suppliers is a major factor in preparing an overall plan of action for any construction process. There is no merit in forging ahead with a plan of construction only to find that it won't work because of delays caused by waiting for materials to arrive on site. Some construction contracts may have a penalty clause (known as liquidated damages). This clause basically says that if the building contractor does not complete the structure and

hand it over to the client by an agreed date, then the building contractor will be fined a financial sum for each day or week it over-runs.

So now we have to consider what information is required to allow the construction operations to succeed, and to do this effectively we need to prepare, and show, all the major construction operations in the form of a bar chart. A bar chart will provide the basis of control for finance, labour, plant and materials.

Before preparing the bar chart the student needs to consider:

- a start date
- what are the sequential operations
- length in time of each sequential operation
- materials requirements and delivery periods

The finish date is ascertained on completion of the set out programme.

To produce the bar chart it is necessary to look at the way in which the information will be shown.

- **time** will be broken down in years, months, weeks and working days
- **sequential** and **non-sequential operations** will be itemised

Time will be shown horizontally in weekly columns broken down into five working days while the sequential and non-sequential operations will be itemised and shown vertically.

The final planning outcome is to produce a series of time/duration bars (Fig. 4.17) running in a horizontal staggered pattern across the sheet, with each bar representing a work operation and showing its length of time/duration true to the time scale used.

Figure 4.17 Bar used in bar chart (see Fig. 4.18), showing start, finish and duration

Under each bar a separate blank space is used as a means of plotting and controlling the actual progress of that work operation on site (see Fig. 4.18).

Self-assessment tasks

1. Using the details given in 'sequence of construction operations for a low-rise dwelling' (p. 170) fill in the 'estimated time in days' column working to the nearest half a day accuracy, e.g. Item 2 = say 4 days, Item 3 = say 0.5 day, etc.
2. Using the sequence details produce a bar chart for the **sequential operations**.
3. Complete the bar chart by adding the remaining **non-sequential operations** to complete contract.

Some useful hints to be considered in producing a programme of works

- The student should remember that the basic concept of a programme is to convey to others the planners' ideas. It

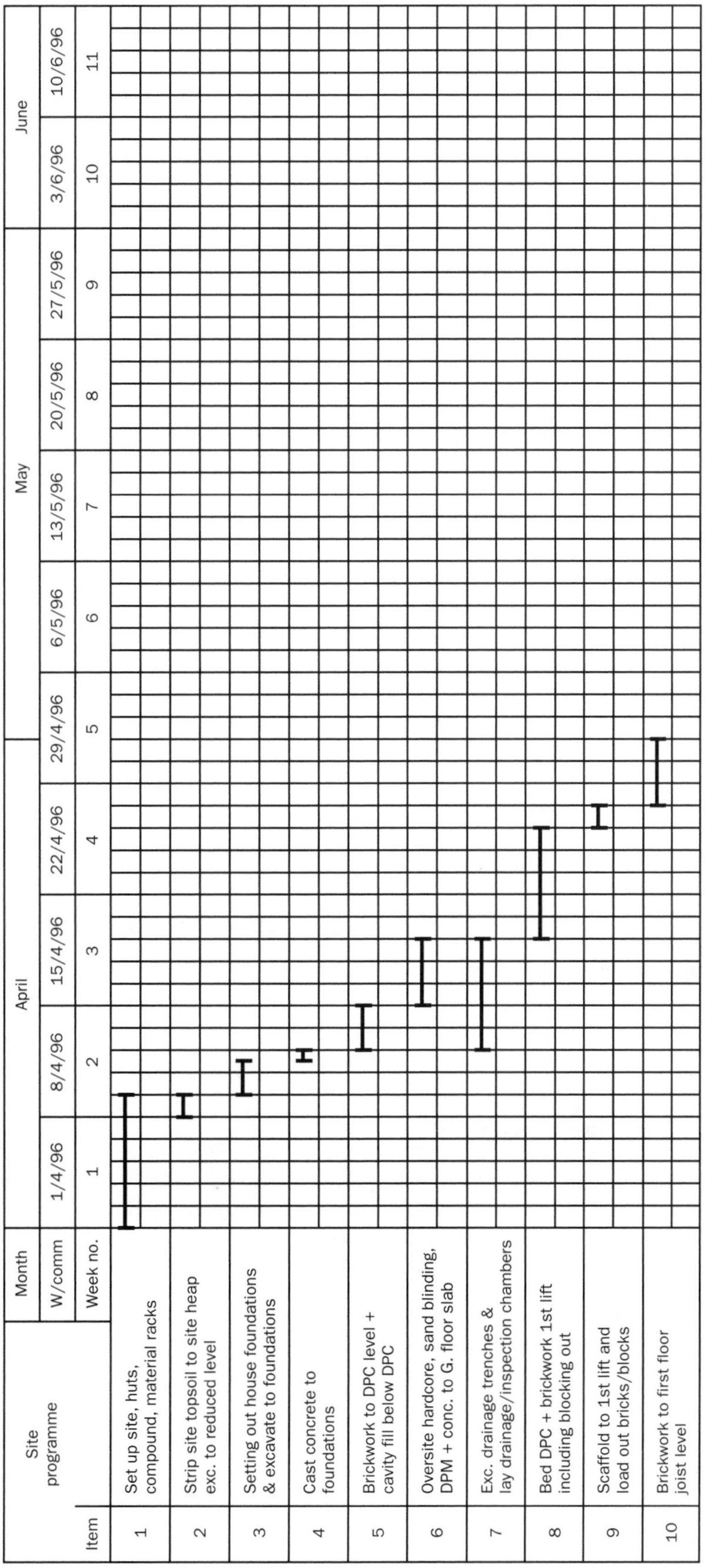

						April				May					June	
	Month		1/4/96	8/4/96	15/4/96	22/4/96	29/4/96	6/5/96	13/5/96	20/5/96	27/5/96	3/6/96	10/6/96
	W/comm												
Item	Site programme	Week no.	1	2	3	4	5	6	7	8	9	10	11
1	Set up site, huts, compound, material racks		▮										
2	Strip site topsoil to site heap exc. to reduced level			▮									
3	Setting out house foundations & excavate to foundations			▮									
4	Cast concrete to foundations				▮								
5	Brickwork to DPC level + cavity fill below DPC				▮								
6	Oversite hardcore, sand blinding, DPM + conc. to G. floor slab					▮							
7	Exc. drainage trenches & lay drainage/inspection chambers				▮								
8	Bed DPC + brickwork 1st lift including blocking out					▮							
9	Scaffold to 1st lift and load out bricks/blocks					▮							
10	Brickwork to first floor joist level						▮						

Figure 4.18 Part of a typical bar chart

will show *what* has to be done, the *order* in which it will proceed; *when* to start the various work operations and *how* long they will last for

- To phase a programme is to simplify its production
- Build the programme around the *sequential* (vital) operations
- Plan as much overlap of work operations as possible; but remember you don't want one trade to get in the way of another
- Keep the number of work operations being carried out at any one time as high as possible
- Plan maximum continuity of work for each of the trades
- Calculating the approximate quantity of work in the *major operations* will show where the greatest efforts should be made, both in planning and in the execution of the work
- An indication of the number of operatives needed to carry out particular work operations in the time allowed on the programme helps those who are responsible for carrying out such work
- The use of notes on a programme helps operatives to understand the purpose of the work
- Aim to get the structure *wind-* and *weather-proof* before 60% of the contract period has expired to allow sufficient time for finishing trades
- Plastering is a vital (sequential) operation and it should be started as soon as the structure is weatherproof
- If you are given a contract period to work to, try to plan completion within 90% of the contract period to allow for any unforeseen circumstances/delays

Setting out planned using trigonometric and algebraic techniques

The process of setting out could be described as methods of reproducing the structures' overall sizes (on plan) on the land area in true relation to surrounding levels and boundaries. It does not matter if the land is level and flat or hilly and uneven because the setting out process relies on fixing suspended data points in the horizontal (truly level 0° horizontally) plane above the ground's surface and transferring them down to the actual land surface. But it does matter that the shape, positioning and sizes are correct.

To transfer down to the land surface the use of plumb will be employed (see Section 4.4, Fig. 4.30). Methods of producing **level** and **plumb** are described under the heading 'techniques used when undertaking basic craft operations'.

A site engineer (usually a surveyor by profession) would be employed on larger projects where the volume of setting out is large enough to keep the engineer fully employed. For smaller projects, say a one-off structure, it is usually carried out by the site agent.

For setting out any building, trigonometric data is used to determine sizes and positions.

Conventions in trigonometric formulas

It is vital that the student understands the use of trigonometrical formulas in the art of setting out of buildings. If you have two items of information then by applying the

formulas or by transposing them you can find the third piece of missing information.

In using trigonometric formulas the following conventions will be used throughout and recognition of them will allow the student to understand the algebraic terms they are always stated in.

- All **angles** are represented by upper-case letters, e.g. A, B or C.
- All **sides** are represented by lower-case letters, e.g. a, b or c.

For example using the formula of Pythagoras' Theorem, $a^2 + c2 = b^2$, tells us that because lower-case letters are involved the sides are being considered here (see Fig. 4.20).

Basic setting out details to be recognised

The following details are used in any setting out operation:

- use of right angles ensure that the building is kept square
- all setting out is to be checked for squareness by measuring the diagonals and adjusting if necessary until they are both equal (Fig. 4.19)
- use of Pythagoras Theorem will form a right-angled triangle $- a^2 + c^2 = b^2$ e.g. a triangle with side ratios of 3, 4, 5.
- use of trigonometrical ratios, e.g. SOHCAHTOA (see below).

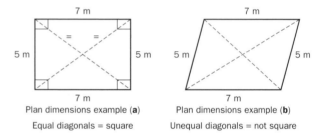

Plan dimensions example (**a**)
Equal diagonals = square

Plan dimensions example (**b**)
Unequal diagonals = not square

Figure 4.19 Examples of correct side lengths for square and parallelogram

Trigonometrical techniques explained

The notation of a right-angled triangle (Fig. 4.20) is related to how the triangle is first set down. Capital letters refer to the angles whereas the lower-case letters refer to the sides.

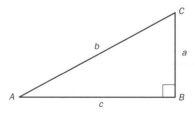

Figure 4.19 Examples of correct side lengths for square and parallelogram

Now consider the sides in greater detail and relate them to their opposite angle in the triangle. The sides can now be referred to in relationship to the angles (see Fig. 4.21).

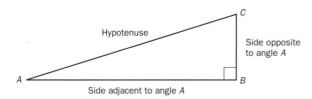

Figure 4.21 Triangles notation considered from Angle A

For angle A:

Hypotenuse: the side from A to C opposite to the right angle

Opposite: the side from B to C opposite the angle A

Adjacent: the side from A to B adjacent to the angle A as it forms the angle A with the hypotenuse

For the angle C (see Fig. 4.22):

Hypotenuse: the side from A to C opposite to the right angle

Opposite: the side from A to B opposite the angle C

Adjacent: the side from B to C adjacent to the angle C as it forms the angle C with the hypotenuse

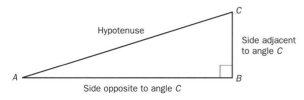

Figure 4.22 Triangle notation considered from angle C

The following trigonometric ratios for any right-angled triangle can easily be remembered by saying the acronym 'SOHCAHTOA'.

The SOH part refers to: **sine** equals **opposite** over the **hypotenuse**

The CAH part refers to: **cosine** equals **adjacent** over the **hypotenuse**

The TOA part refers to: **tangent** equals **opposite** over the **adjacent**

The above three examples can be expressed as:

$$\text{Sine} = \frac{\text{opposite}}{\text{hypotenuse}}$$

$$\text{Cosine} = \frac{\text{adjacent}}{\text{hypotenuse}}$$

$$\text{Tangent} = \frac{\text{opposite}}{\text{adjacent}}$$

Trigonometrical functions for right-angled triangles used in setting out

The at-a-glance formula sheet in Fig. 4.23 shows the formulas required for the differing cases the student is likely to require/select during calculation stages.

Example

For the triangle shown below, calculate (a) the value of the angle A, and (b) the length of side a.

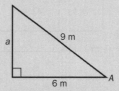

Answer

(a) We know the lengths of the side adjacent to the angle B, and of the hypotenuse. Using SOH**CAH**TOA, we can say that the cosine of A is given by:

$$\cos A = c/b = 6/9 = 0.667$$

Using the inverse cosine ($\cos^{-1}$) function of a calculator, or a book of tables, we can see that:

$$A = \cos^{-1} 0.667 = 48.2°$$

(b) For a right-angled triangle, we can use Pythagoras' theorem, $b^2 = a^2 + c^2$. We know b and c, and hence we can find a as follows:

$$b^2 = c^2 + a^2$$
$$9^2 = 6^2 + a^2$$
$$81 = 36 + a^2$$
$$a^2 = 81 - 36 = 45$$
$$a = \sqrt{45} = 6.7\,\text{m}$$

Alternatively, since we know angle A and the length of the hypotenuse b, we can calculate the length of side a by trigonometry. Side a is opposite angle A, so we can use the formula for sin A:

$$\sin A = a/b$$
$$\sin 48.2° = a/9$$
$$0.745 = a/9$$
$$a = 9 \times 0.745 = 6.7\,\text{m}$$

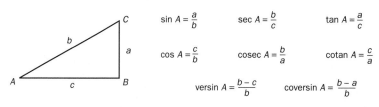

GIVEN	REQUIRED	FORMULAE		
a, b	A, C, c	$\sin A = \dfrac{a}{b}$	$\cos C = \dfrac{a}{b}$	$c = \sqrt{(b+a)(b-a)}$
a, c	A, C, b	$\tan A = \dfrac{a}{c}$	$\cot an\, C = \dfrac{a}{c}$	$b = \sqrt{(a^2 + c^2)}$
A, a	C, c, b	$C = 90° - A$	$c = a \times \cot an\, A$	$b = \dfrac{a}{\sin A}$
A, b	C, a, c	$C = 90° - A$	$a = b \times \sin A$	$c = b \times \cos A$
A, c	C, a, b	$C = 90° - A$	$a = c \times \tan A$	$b = \dfrac{c}{\cos A}$

Figure 4.23 Trigonometrical functions for right-angled triangles

4.4 Basic craft operations

Topics covered in this section are:

- Basic craft safety procedures
- Selecting materials and equipment appropriate to basic craft operations
- Preparing a schedule of materials and equipment
- Investigating techniques used when undertaking basic craft operations

Safety procedures

It should be noted that all tools (whether they be hand or power tools) should be kept in efficient working order. For any tool to do the job of work it is designed to do it should be maintained and kept in a good condition at all times. If it is a cutting tool then the cutting edge should be continually sharpened and protected from damage during storage in the tool bag or tool box.

Just as it is important that the craftsperson selects the correct tool for the work operation in hand, it is equally important that the craftsperson selects and uses correct Personal Protective Equipment for that operation.

Where tools have a cutting edge this should be kept sharp at all times to minimise the likelihood of industrial injury while using them. Blunt tools make heavy going of any work operation as additional physical effort and control is required thus increasing the likelihood of accidents.

Using cutting tools

To minimise accidents using cutting tools general rules should be observed at all times during their use. To do this it is necessary to consider and plan the work operation before the task is carried out and to enforce the following general rules (which apply to all crafts):

- Keep both hands and body behind the cutting edge at all times in order to give complete control of the tool. This reduces the risk of personal injury should the tool slip.
- Body posture, e.g. balance and the correct level to work at, will ensure complete control of all physical effort in using the tool.
- Sawing should be an effortless, controlled, easy process of moving the arm back and forth. If extra physical effort is required to make it cut, then the saw should be sharpened.

Selecting materials and equipment appropriate for basic craft operations

Each craft's suggested tool kit listed here is not the definitive collection. Each craftsperson will collect/make/purchase additional tools throughout their working life and possibly well into retirement. What is listed are the main tools required for each trade.

Apprentice carpenter's tool kit/equipment

- Tool bag or box
- Square and sliding bevel
- Screwdrivers
- Rule
- Saws – rip, panel, tenon, coping
- Chisels – 6 mm, 9 mm, 12 mm, 15 mm, 19 mm, 25 mm, 32 mm
- Steel smoothing plane
- 20 oz claw hammer, club hammer, Warrington hammer
- Combination oil stone
- Oil can (very light oil)
- Ratchet brace
- 6 mm, 9 mm, 12 mm, 15 mm, 19 mm, 25 mm, 32 mm auger or Jennings Patt. bits
- Steel jack plane
- Nail punches, large and small
- Hand drill (wheel brace) and bits
- 1 m spirit level
- Plumb bob and line
- Pincers
- Pliers
- Wooden mallet
- 50 mm bolster
- Bradawl
- Hacksaw

Apprentice bricklayer's tool kit/equipment

- Tool bag or box
- Straight edge 600 mm long
- 250 mm bricklaying trowel
- 100 mm bolster chisel
- 1400 mm spirit level (plumb and level bubbles)
- 2.5 lb club hammer, scutch hammer
- 1800 mm steel tape
- Pins and lines
- Corner blocks and lines
- a range of cold chisels
- 38 mm comb chisel
- Soft broom head
- 150 mm pointing trowel
- 175 mm square pointing hawk
- Frenchman
- Boat level
- 175 mm pincers
- Screwdriver
- Jointing irons

Apprentice plasterer's tool kit/equipment

- Tool bag or box
- 450 mm or 600 mm level (plumb and level bubbles)
- Laying-on trowels (floating and finishing)

- 200 mm gauging trowel
- Internal and external trowels
- Pointing trowel
- Alloy hawk
- Floor-laying trowel
- 150 mm wetting brush
- Margin trowel
- 1800 mm steel tape
- Wooden floats: Scratching 300 mm × 125 mm × 12 mm
  Skimming 300 mm × 125 mm × 12 mm
  Small 200 mm × 100 mm × 9 mm
- Cold chisels and bolster
- 2.5 lb club hammer
- Lath hammer
- Tin snips
- Pincers
- Square
- Sharp retractable blade knife
- Saw
- Chalk line
- Straight edges: Floating rule 150 mm × 100 mm × 19 mm
  Feather edge 125 mm × 100 mm × 15 mm
  (chamfered)
- Scoop or small shovel (stone dashing)
- Wire drag scratcher

Apprentice plumber's tool kit/equipment

- Tool bag or box
- Large hacksaw
- Copper tube cutters, 12 mm–32 mm I/D
- Dresser
- 250 mm tin snips
- Screwdrivers
- 175 mm and 300 mm pipe wrenches
- Bending springs
- Cold chisel and bolster
- 2.5 lb club hammer
- Steel tape
- Files
- Bossing stick
- Lead knife
- Shave hook
- Turnpin (small)
- Wiping cloths, 38 mm, 62 mm, 90 mm widths
- 750 mm level (plumb and level bubbles)
- Boat level
- Chasing chisel
- Ball pein hammer
- Caulking irons (3 sizes)
- Ratchet brace
- Twist bits, 19 mm and 32 mm
- Wheelbrace (hand/breast drill)
- Timber mallet (medium size)
- General purpose saw
- Pincers and pliers
- Line
- Gas torch equipment
- 12 mm and 19 mm wrench (bath and basin taps)

Apprentice painter/decorator's tool kit/equipment

- Tool bag or box
- Straight edge 600 mm long

- Knives – stripper, stopping, putty, hacking, palette
- Shave hook, pear shaped
- Shave hook, diamond shaped
- Dusting brush
- Wire brush
- Paper hanging shears
- Paper hanging brush
- Plumb bob and line
- 1 m folding boxwood rule
- Wooden roller (paper hanging)
- Claw hammer
- Nail punch

Self-assessment tasks

Using your library facilities identify the tools required together with a brief explanation of the use of each tool in:

1. Apprentice carpenter's tool kit
2. Apprentice bricklayer's tool kit
3. Apprentice plasterer's tool kit
4. Apprentice plumber's tool kit
5. Apprentice painter and decorator's tool kit

Preparing a schedule of equipment

A schedule of equipment (Fig. 4.24) can be produced for any work/job operation listing the type of equipment and periods for which it will be required.

Types of tools

Some of the following listed tools may be used by more than one specific trade for the finishes they produce.

Saws

Learn to recognise the different types of saw teeth and why they are so shaped. The types of saw teeth are:

- cross cut
- rip

Get to know the differences between them.

Sizes of saw teeth

Dictated by types of saws which are designed for specific cutting purposes, teeth may be:

- fine cut
- coarse cut

Why do they vary in size?

Additional saw teeth requirements

Consider also why saws have different widths of cut for efficient operation. What requirement assists this?

- saw teeth set

```
┌─────────────────────────────────────────────────────────────────┐
│  PLANT/EQUIPMENT SCHEDULE                                         │
│                                                                   │
│  CONTRACT: MR NEWALL                              No. 007         │
│  JOB No. 220658                                                   │
│  PREPARED BY: AP                              DATE: 03-FEB-92     │
│                                                                   │
```

PLANT TYPE & EQUIPMENT	DATES REQUIRED		EQUIPMENT TO BE SUPPLIED BY	REMARKS
	DATE ON SITE	DATE OFF SITE		
PORTABLE SAW	1-APR-92	31-OCT-95	HEAD OFFICE STORES	LONG TERM
4 ACROW PROPS	4-MAY-92	19-DEC-92	TO BE HIRED	
8 WORKMATES	1-APR-92	31-OCT-95	ALREADY ON SITE	OLD CONTRACT

Figure 4.24 A typical plant/equipment schedule

Self-assessment tasks

1. Sketch the differences between cross cut and rip saw teeth and state why these differences exist.
2. Explain the difference between coarse and fine saw teeth and how they are specified.
3. Using a neat annotated sketch show why the 'set' to a saw's teeth is vital to its efficient cutting operation.

Saw types (hand use)

Panel saw – cross cut and rip
Backed saw – tenon
Pad saw
Coping saw
Hack saw
Masonry saw
Jigsaw
Portable circular saw
Chain saw
Reciprocating saw

Saw types (workshop fixed)

Band saw
Bench jigsaw
Radial arm saw
Circular saw bench

Planes

Block plane
Smoothing plane
Jack plane
Bull nose plane
Bench rebate plane
Shoulder plane
Plough plane
Combination plane

Routers

Hand router
Portable electric router

Chisels

Chisel types differ according to the type of work they are designed to do:

- angles at tip (grinding bevel)
- cutting tip (sharpening bevel)

Chisel types

Firmer chisel
Paring chisel
Mortice chisel
Electrician's floor board chisel
Firmer and paring gouges
Glazier's chisel
Brick bolster chisel
Cold chisel
Carving chisels and gouges

Holding devices

Woodworker's vice
Metalworker's vice
Workmate bench
'G' cramps
Bar cramp

Sharpening tools

Bench grinder
Oilstone
Saw set

Marking out tools

Straight edge
Steel rule
Try square
Sliding bevel
Mitre square
Dovetail template
Glazier's 'T' square
Marking gauge
Mortice gauge

Setting out tools

Straight edge
Steel rule
Steel tapes (various lengths)
Folding rules
Spirit levels
Carpenter's steel square
Chalk line
Plumb bob
Bricklayer's line and pins

Finishing tools

Spokeshave
Double handed scraper
Metal scraper
Painter's shave hook
Painter's paint scraper
Painter's window scraper
Smoothing planes
Glasspaper
Garnet paper
Emery paper/cloth
Aluminium oxide paper/cloth
Silicon carbide paper (wet and dry)
Tungsten carbide discs
Glasspaper block
Disc sander pad
Belt sander
Finishing sander

- It is good practice to ensure that all finished materials have the arris (sharp edge) removed to prevent cuts or splinters to those handling the materials.
- Timber being prepared for painting should be sanded with a medium grade glasspaper rubbed at an angle of 45 degrees to the sides to provide a roughened key for primer, i.e. across the grain.

Stripping tools

Blow lamp
Painter's shave hook
Painter's paint scraper
Painter's window scraper

Cutting tools

Saws (various types)
Chisels (various types)

Knives – trimming, glazier's, decorator's
Decorator's wallpaper trimmer
Rasps
Files
Tin snips
Paper hanger's scissors
Thread cutting taps and dies
Plumber's pipe and tube cutter

Application tools

Glazier's putty knife
Filling knife
Palette knife
Tube bender
Soldering iron
Paint brushes
Paint roller
Paste brush/wall brush
Washing down brush
Paint pad
Mottler brush
Pencil overgrainer brush
Flogger brush
Softener brush
Stippling brush
Stencilling brush
Lining fitch brush
Fitch brush
Radiator brush
Paper hanger's brush
Seam roller
Plasterer's steel float
Wooden float (bricklayer and plasterer)
Plasterer's trowels
Flooring trowel
Edging trowel
Bricklayer's pointing trowel
Bricklayer's jointer

Driving tools

Claw hammer
Pin hammer
Cross pein hammer
Engineer's hammer
Club hammer
Sledge hammer
Brick hammer
Joiner's mallet
Carpenter's cabinet screwdriver
Electrician's screwdriver
Spiral ratchet screwdriver
Ratchet screwdriver
Stubby screwdriver
Nail punch

Extraction tools

Pliers
Mole grips
Glazier's pliers

Carpenter's pincers
Crowbar

Boring tools

Hand brace and bits
Wheel brace
Breast drill
Hand power drill
Carbon steel 'Morse' twist drill bits – suitable for wood only
High speed steel 'Morse' twist drill bits – suitable for wood and metal

Turning tools

Spanners (open ended, ring, offset ring, multiple ring, split ring, combination and adjustable)
Monkey wrench
Centre screw wrench
Crescent spanner
Box spanner
Socket spanner
Allen keys
Pipe tong wrench
Stilson wrench
Chain wrench

Support tools

Tool bag
Tool box
Decorator's pasting table
Electricians's/plumber's pipe bender

Setting out for craft operations

Before any item can be manufactured it is necessary to set out the construction details to enable the workshop operative to work to exact details. These details will be drawn to a recognised scale or to full size. Full size details are preferred for workshop 'rods'. The word rod is a trade term used to describe any setting out.

In shopfitting, for example, it is impossible to set out to full size the lengths and heights of a shopfront and the interior on paper. Therefore a system of reducing the measurement lengths is used whereby the shopfront is condensed to just showing the relevant junctions and section/fixing details with exact figured dimensions added to assist the workshop operations and also the fixer when it is eventually sent out to site.

Preparing a schedule of materials

In order to produce a schedule of materials (Figs 4.25 and 4.26) you require to know the length, width, thickness and final shape of all material taken from the workshop rod. The examples given here are based on the joiner's workshop rod details for the manufacture of a purpose-made softwood window frame.

Notes on reading materials lists

It should be realised by the student that all works associated in reading a materials list are related to, and should be read with, the setting out drawings; in this case drawing numbers 240796/1A Window Elevation and 240796/2A Window Sections (Fig. 4.25). You may also come across the term 'Cutting sheet' especially when attending a wood machining workshop. This is the same as the materials list but is the term used by wood machinists because of their role in getting the timber out. They have no interest in ironmongery, etc., and therefore these items are left off for them.

How to read the materials list and the column headings

Item numbers

These are used to identify the number of separate items required.

Mark numbers

These are given to each timber section to allow the wood machinist to identify and machine the timber sections to the correct size and shape.

All Mark Numbers will be marked on the timber by a 'Marker-Out' (craftsperson with responsibility for marking out the timber sections).

Mark numbers also assist the joiner by making the job of identifying timber sections easier when reading the setting out drawings and putting the frame together.

Description

Describes the item to assist easier recognition.

No. off

This is the trades description (which you may not have come across so far) and means 'number of pieces required' for this particular item, e.g. for the wood machinist to get out and the joiner to find when constructing the frame.

Nominal sizes

Nominal sizes are not actual finished sizes, but the sizes used to order the material, e.g. material in the sawn state as supplied. It is recognised that the nominal size of timber sections is a minimum of 3 mm thicker than the finished dimensions.

Nominal sizes refer to the:

- overall length of the timber section
- width and thickness, which give the timber cross section size

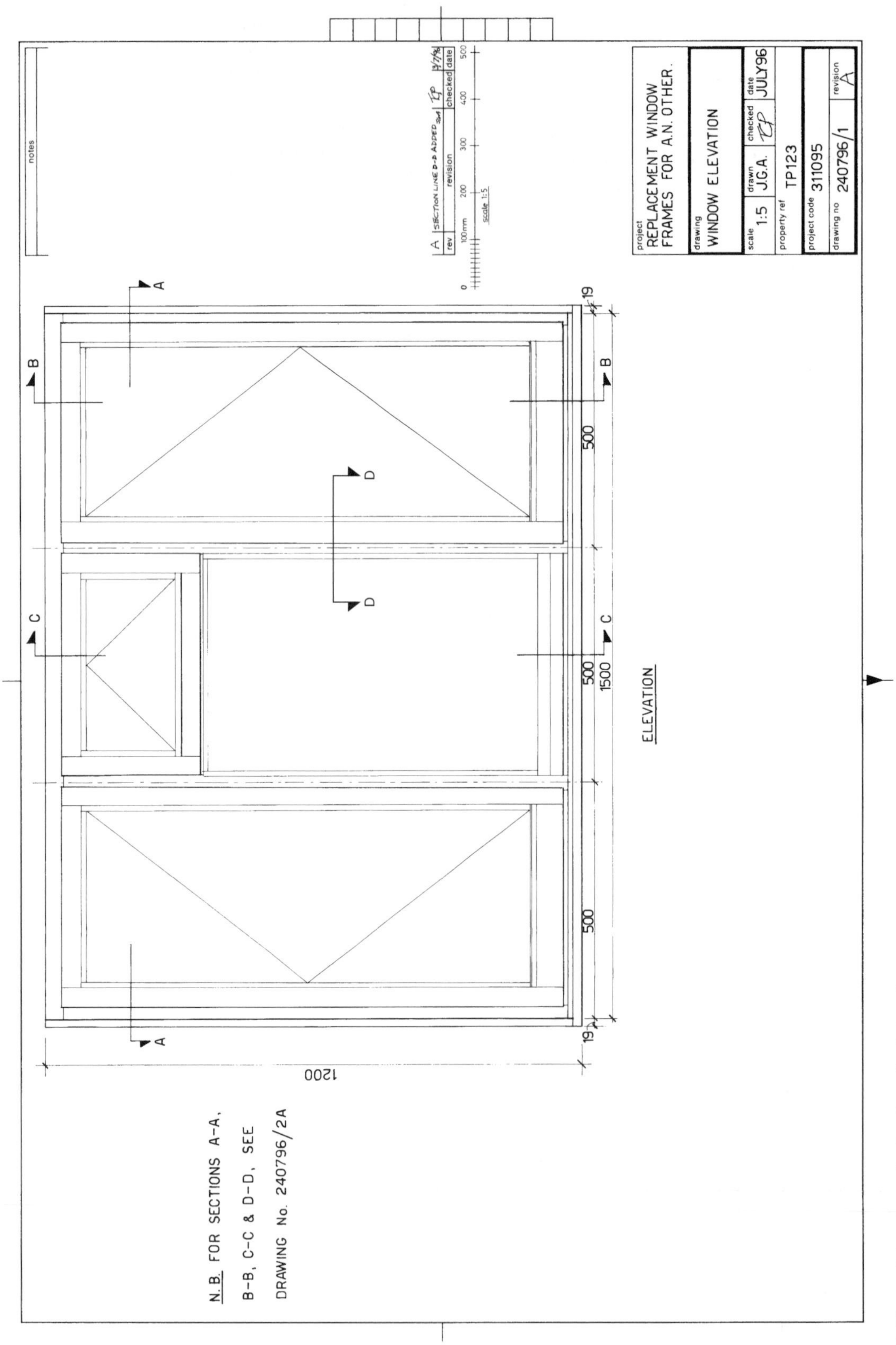

Figure 4.25 Setting out drawings of a hardwood window frame

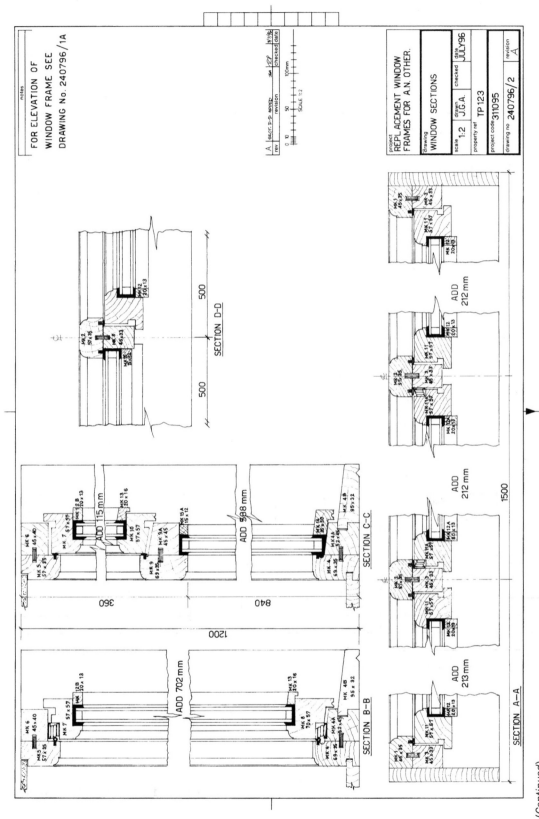

Figure 4.25 (Continued)

Materials list

Item	Mark No.	Description	No. Off	Nominal sizes			Finished sizes		Remarks
				Length	Width	Thickness	Width	Thickness	
1	1	Jamb	2	1220	50	38	45	35	Softwood
2	2	Mullion	2	1220	75	38	57	35	Softwood
3	3	Mullion/Jamb	4	1220	50	38	45	33	Softwood
4	4	Cill	1	1520	75	38	69	35	Softwood
5	4a	Cill	1	1520	62	50	52	45	Softwood
6	4b	Cill	1	1520	100	38	95	32	Softwood
7	5	Head	1	1520	62	38	57	35	Softwood
8	6	Head	1	1520	50	50	45	40	Softwood
9	7	Window top rail	3	480	62	62	57	57	Softwood
10	8	Window bottom rail	2	480	75	62	70	57	Softwood
11	9	Transom	1	550	75	38	69	35	Softwood
12	9a	Transom	1	550	50	50	45	45	Softwood
13	10	Sash bottom rail	1	480	62	62	57	57	Softwood
14	11	Sash styles	4	1200	62	62	57	57	Softwood
15	11a	Sash styles	2	300	62	62	57	57	Softwood
16	12	Glazing bead	4	1050	25	20	20	13	Softwood
17	12a	Glazing bead	2	200	25	20	20	13	Softwood
18	12b	Glazing bead	3	900	25	20	20	13	Softwood
19	13	Glazing bead	3	400	25	20	20	16	Softwood
20	14	Glazing bead	1	450	38	38	35	30	Softwood
21	15	Glazing bead	2	810	25	20	15	12	Softwood
22	15a	Glazing bead	1	480	25		15	12	Softwood
23		Loose ply tongues		12 Lin.M			17	6	Birch plywood
24		601 series HD friction	2 pairs	350					
		Window stay hinge							
25		201 series HD friction	1 pair	200					
		Window stay hinge							
26		Code 0305 monostrip		8.120 Lin.M					
		Draught excluder							
27		Double glazed sealed	2	1080			365	20	
		Unit							
28		Double glazed sealed	1	794			460	20	
		Unit							
29		Double glazed sealed	1	197			383	20	
		Unit							

Figure 4.26 Schedule of materials for window in Fig. 4.25

Finished sizes

The workshop drawing showing the sections through will be the reference for the finished sizes.

All finished sizes are, as the term implies, the actual final cross section sizes that the wood machinist would get the timber out to.

Remarks

This column is used for adding notes and diagrams relating to the material item. It should be used where information is necessary to allow the reader to identify details that otherwise cannot be easily recognised.

Techniques used in basic craft operations

Before the nineteenth century the main construction material in this country was timber (English Oak) and little attention was given by carpenters to line, level and plumb; hence the timber framed structures of this period which are still standing may seem to be poised at a peculiar set of angles with no attention being given to forming right angles in the horizontal or vertical plane (line, level and plumb).

Remember, a modern craftsperson will work to line, level and plumb whenever carrying out a craft operation.

Methods of producing a line

A line is produced by using one of the following:

- straight edge
- spirit level
- string line
- steel rule

A craftsperson will check the accuracy of a straight edge in the following way:

1. Scribe a line from the straight edge.
2. Roll over the straight edge 180 degrees to align the straight edge with the previously scribed line.
3. If a difference appears as shown in Fig. 4.27, then it proves that the straight edge is not completely accurate.
4. But if the line is the same as previously scribed in 1, then it is completely straight.
5. Another method is for the craftsperson to pick up one end of the straight edge and sight an eye down the length of the edge to be used. If seen to be bowed it would either be checked and corrected as 1, 2, 3 and 4 above, or discarded.

Self-assessment tasks

1. Select a piece of timber and check it for straightness using the steps 1 to 4 shown above.
2. Cast an eye down the length of material and see if you can observe the shape as confirmed in task 1.

A useful general straight edge is 2400 mm long but shorter tailor made lengths may be required for specific work operations.

A typical straight edge (Fig. 4.28) should have the following characteristics:

- be made from any stable material e.g. plywood, steel or plastic sheet material, etc.
- have a thickness to afford rigidity/stability (preventing whipping) when picked up and in use
- both longitudinal edges truly parallel so that a spirit level can be applied when working to level or plumb elevations/surfaces

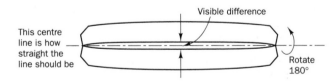

Figure 4.27 Checking the accuracy of a straight edge

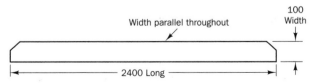

Figure 4.28 A typical plywood straight edge for use in a workshop or on site

Level and plumb

Level can be described as being truly horizontal or zero degrees in the horizontal plane (line A–B in Fig. 4.29).

Figure 4.29 Example of level

Plumb can be described as being truly vertical or as a right angle or 90 degrees to the horizontal plane (line C–D in Fig. 4.30).

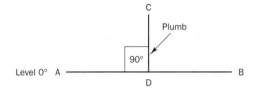

Figure 4.30 Example of plumb

Methods of finding level

The use of the following items will give a craftsperson a level line or surface:

- spirit level
- water level

A spirit level (Fig. 4.31) is basically a straight edge housing a bubble in the centre of its length. The bubble is set to give a reading in a centre position when the level is exactly level. The craftsperson then adjusts the work to the centre position reading.

Figure 4.31 A typical spirit level

A water level (Fig. 4.32) consists of a pair of reading tubes connected by a length of tubing. It works on the principle of physics that when both tubes are unsealed allowing in atmospheric pressure the water in the tubes and tubing will always rise or fall to find its own level of equilibrium.

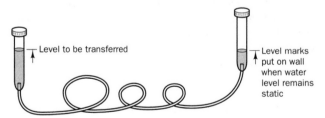

Figure 4.32 A water level

Providing the pair of tubes is kept static and the connecting tubing is not restricted by kinks or air locks then equilibrium of the water level will give levelling points. The water level is particularly useful where you cannot see the other levelling point, such as transferring levels from room to room with walls in the way. The only limit on its use is the length of connecting tubing.

Methods of finding plumb

The use of the following items will provide a craftsperson with a plumb line or surface:

- spirit level with vertical bubbles
- plumb bob
- plumb rule

A spirit level with three or more bubbles is particularly useful as it usually contains a pair of level and plumb bubbles. The example shown in Fig. 4.33 is of an aluminium lightweight level. Spirit levels can also be used as a straight edge.

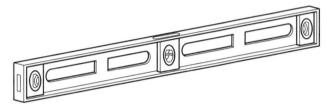

Figure 4.33 A lightweight combination spirit level

A plumb bob (Fig. 4.34) consists of a heavy weight attached to a length of cord. When supported on the cord the plumb bob will eventually stop swinging to give a plumb line.

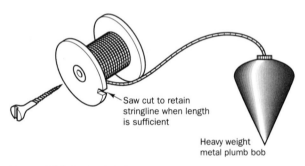

Figure 4.34 A plumb bob

A plumb rule (Fig. 4.35) could be described as a combination of a straight edge and a plumb bob. A straight edge of parallel width (and of suitable length for the work in hand) is marked with an accurate centre line throughout its length. A hole is cut at the bottom of the centre line to accommodate the bob while a single saw cut is produced at the top in the centre line. The cord of the plumb bob is attached and wedged in the top saw cut allowing the suspended bob weight to be adjusted until it is central and free to move in the bottom hole. With the whole apparatus erected vertically and placed against a vertical surface it is said to be plumb when the cord (and the static bob) line up by sight with the centre line.

How to check a spirit level for accuracy

Any spirit level (Figs 4.31 and 4.33) can from time to time become inaccurate. This inaccuracy is mainly brought about by rough handling, e.g. accidental dropping. Providing periodical checks are carried out the inaccurate usage can be reduced to a minimum. The check is best carried out as follows:

1. Put two screws (hereafter referred to as the support screws) in a timber surface a suitable distance apart to support the level by its extreme ends. Make sure the support screws are as near vertical as possible.

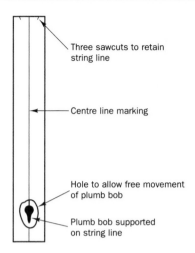

Figure 4.35 A plumb rule

2. Adjust the support screws so that they appear both the same in height.
3. Place the spirit level on top of the support screws and observe if the bubble is reading truly level (i.e. in the middle of the tube). It is unlikely that the support screws have been set accurately enough to read level at this stage.
4. Check to decide which end (right hand or left hand) support screw needs to be adjusted (up or down) to get the bubble level and observe by how much the support screw needs to be adjusted.
5. Adjust the support screw until the level reading is obtained.
6. With the support screws set (assumed level at this stage because it is assumed that the level is reading correctly) the level from this point on is going to be checked for its accuracy.
7. Lift and rotate the level horizontally (end for end) through 180° and place back on the support screws. Note the bubble reading. If the level is accurate the bubble will be in the centre. If the bubble is off centre then it proves that the level is inaccurate and steps 8 to 10 for adjustment must be followed.

How to adjust an inaccurate spirit level

8. Leaving the level in position as described in 7 above, unscrew the clamp holding the bubble tube and carefully rotate until the bubble moves half the distance between the truly level and the previous reading; re-tighten the clamp.
9. Adjust a support screw so that the bubble is reading dead level.
10. Repeat item 7 above.

Note: if the level is still requiring adjustment then items 8 to 10 should be repeated until corrected.

Basic craft operation: bricklayer

Bricklaying, like all crafts, looks easy when you observe a bricklayer at work. Students carrying out bricklaying operations will have the chance of finding out for themselves just how these skills have to be learned by practice.

Brick dimensions

Architects work in co-ordinated brick sizes when designing a building. Each brick that is laid has a specific known dimension for the bricklayer to work to. To consider this we need to look at the sizes of a brick as shown in Fig. 4.36 and then to look at what makes up the co-ordinating brick size.

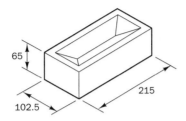

Figure 4.36 Typical brick dimensions in mm

When a wall is constructed by a bricklayer the bricks are bonded together with a mortar (consisting of cement powder, lime, sand and water mixture) which eventually sets to provide a strong element of construction.

Once a brick is laid it forms a module of known dimensions which take into account the mortar joint thicknesses. Bricklayers always work to these modules known as the **co-ordinating size** and **brick gauge** (see Fig. 4.37). The mortar joint is 10 mm thick when laid for each horizontal and vertical joint, referred to as bed and perp. joint respectively.

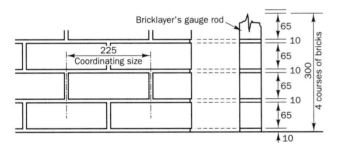

Figure 4.37 Guage rod and brick dimensions when laid

We have already mentioned that the word 'rod' is a craft term describing any setting out, and the bricklayer is no exception to this as the brick courses have to be planned and known before bricklaying begins. A vertical 'gauge rod' is set out to ensure that each course is laid at the right distance apart and is marked out on a length of timber in increments of 75 mm, which is the height of each brick gauge.

A 'storey rod' will also be marked out with the brick course (gauge) increments but with additional information added in the form of the heights where other construction elements would be built into the brickwork, e.g. window cills, lintels, first floor joists, etc.

Selection of bricks from brick packs

Labourers are employed by bricklaying gangs to mix and transport the mortar and load-out the workface with bricks/blocks for the bricklayer to lay. The selection of facing bricks by the labourer is very important as brick packs have variations in colours. It is not uncommon to see buildings where bands of different non-contrasting bricks can spoil the aesthetics of a structure. To prevent this the labourer should take bricks from as many packs as possible (minimum of three) in order to blend them.

Spot boards

Spot boards are usually plywood boards positioned approximately 600 mm from the working face (the wall to be built) and are used for the containment of the mortar deposited by the labourer. They alternate with stacks of bricks along the face of the wall to be built. The labourer's job is to make sure that each bricklayer is constantly fed sufficient mortar for the work in hand.

Setting out the brickwork before laying (cavity walling)

Prior to any work starting on setting out the brickwork the building size will have already been set out by the site engineer/agent by erecting timber profiles (Fig. 4.38) with the brickwork data/sizes marked on the top edge of the profile boards and located by nails driven in for the string lines.

Once the foundation concrete is cast and set the bricklayer can begin the task of setting out the building by marking the extreme corners in fresh mortar thinly spread on the concrete foundation's top surface.

To do this it is necessary to attach string lines to the nails marking the outside edge of brickwork. Then check and adjust for square (by measuring the diagonals) just in case the profile boards have been knocked during the foundation trenching and casting process. Where the string lines cross at the corners the position has to be plumbed down using a spirit level (or plumb bob or plumb rule) to mark the extreme corners of the outside skin of brickwork. The outside skin, being the overall size of any structure, is all that is required to be set out to enable the bricklaying to begin. The internal skin is secondary at this stage although the width of the cavity walling will also be marked at the same time as the outer by moving the string lines to the inner nail position.

Some bricklayers prefer a dummy run by laying the first course of bricks dry after marking out the corners (and before eventually laying them in mortar) along the line of each side in order to get the feel of the course and work out the brick co-ordinating sizes. Once the bricklayer is sure of the work then all dry laid bricks are removed and the actual bricklaying operation gets underway using the cement mortar for bed and perp. joints.

Basic craft operation: carpenter and joiner

The basic difference between carpentry and joinery is that the carpenter is mainly involved with works on site:

- shuttering
- floor joists
- first and second fixing
- roof construction

The joiner works mainly in a workshop producing items of joinery in hardwood and softwood:

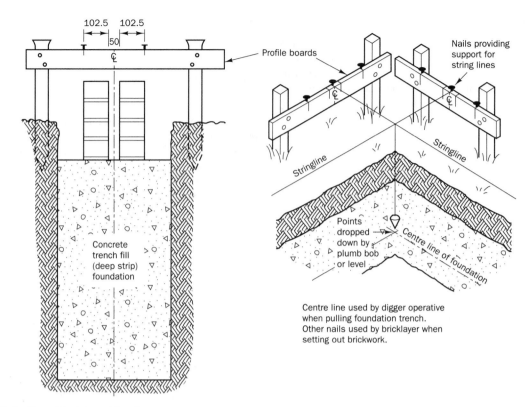

Figure 4.38 Profiles with data and location on corners

- window frames
- door frames
- staircases
- internal fittings

In shopfitting the joiner will do some metalworking as well.

Using and working timber

Timber, being a fibrous material, has to be treated with certain respect when employing tools to work it. It is not uncommon for joints near the end of the material to be split open by chopping out too much material using blunt tools instead of taking small bites with a sharpened tool. Sharp tools leave a neat cut finish whereas blunt tools tend to rip out leaving uneven and damaged surfaces.

Using a plane on timber calls for certain rules to be obeyed. Because timber is made up of fibrous cells it has a grain. The grain can be observed by looking at the annual rings running through the timber and usually running out at the edge. It is not good practice to plane timber against the grain as splitting out will occur. Therefore all hand planing should be carried out going with the direction of the grain, leaving a smooth finish. This grain can actually be felt on sawn timber by carefully and lightly running the fingers slowly backwards and forwards. One way the grain lies flat and it feels fairly smooth; the opposite way the grain will rise up and can be felt by the sensitivity of the finger tips. Extreme care has to be taken when doing this so that splinters are not picked up in the fingers.

Timber is always ordered on a cutting sheet at least 3 mm in section oversize to allow for preparation down to the exact size required. In other words, when the timber mill gets the material out it will be in a sawn state initially. Preparation in

this case means planing the timber all round on all four faces to a smooth surface.

When a joiner prepares timber by hand:

- the material is always in the sawn state to start off with, i.e. straight from the circular saw in the mill
- it has to be physically prepared using a jack plane

Before any planing operation takes place the joiner has to ascertain two vital things for each piece of timber selected:

- which side is going to be **face side** on each piece
- which edge is going to be the adjacent **face edge**

The reason for making the selection of face side and face edge (see Fig. 4.39) is to ensure uniformity because all face sides will face upward and all face edges will face inwards. This ensures two things: firstly, that any discrepancy in the timber thickness does not affect the alignment of the joints because the face sides will all be flush with one another; and secondly, that each joint shoulder is fitted square with the face edge. Should any discrepancies appear in the joint thickness then these are dealt with in the cleaning-up stage by planing the rear of the joints to bring the material to a constant thickness. The face side should only require a fine shaving to clean it up ready for sanding.

Remember:

- any marking out or gauging in the thickness of the timber must be carried out from the face side
- any marking out or gauging in the width of the timber must be carried out from the face edge.

The angle formed by the face side and face edge must be exactly a right angle (90 degrees) and is constantly checked by offering a 'try square' against the edges when the material is being planed up. The face side is prepared first and checked for complete

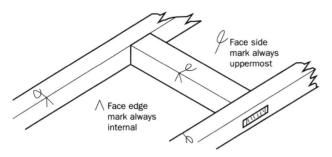

Figure 4.39 Example of the use of face side and face edge marks

flatness; then the face edge is planed up and checked for square using a 'try square' with the stock placed against the face side of the timber and reading the squareness on the underside of the blade. The next step is to gauge the rear side and edge and to plane them to the final required width and thickness.

By now it should be evident that a set of disciplines have to be followed in order to produce a quality item of joinery. No matter which trade we look at these disciplines still exist and the quality of the outcome of the work is a measure of the craftsperson's skills.

Operatives, where comments can be made, are numbered 1 to 14.

Operative No. 1
- Standing on a hop-up which contravenes the regulations and is totally unsafe
- Is over-reaching for laying bricks

Operative No. 2
- Operating a circular saw with crown guard not in safety position
- Cutting timber freehand – no fence being used to run parallel width timber
- Fingers too near saw blade – push stick not evident
- A saw bench this size would almost certainly be powered by a three-phase 415 volt electricity supply and should be sited in a temporary site saw shed with correct electrical installation, not out in the open as shown – nothing to stop other operatives falling on to it
- Bad housekeeping – offcuts/rubbish lying around and not cleaned up
- Bad housekeeping – dangerous stacks of loose materials precariously poised to become unstable

Operative No. 3
- Using the scaffold which is obviously incomplete. 'INCOMPLETE SCAFFOLD – DO NOT USE' warning notice not in evidence to warn operatives of the impending dangers

Operative No. 4
- Dangerous practice of forming a makeshift hop-up on a scaffold platform – trip hazard for other users
- Also, unsafe storage of materials on hop-up – may tip up when operative gets off

Operative No. 5
- Operative contravening regulations by riding on a hoist *not* designed to carry operatives

Operative No. 6
- Over-reaching to grab bucket on gin wheel because no safety rail is fitted to scaffold

Operative No. 7
- Carrying materials with restricted view, especially to the left

Operative No. 8
- Unsafe carrying of a ladder – about to make contact with operative No. 7 who won't know what hit him

Operative No. 9
- Unaware of close proximity of backactor bucket
- Working with back to machine
- Accident waiting to happen if digger driver has limited vision with operative in a blind spot

Operative No. 10
- Strictly illegal to ride on lifting operations

Operative No. 11
- Is not wearing a safety helmet

- Could be struck by pipe in transit to trench
- Pickaxe on side could easily be accidentally kicked in

Operative No. 12
- Unsafe practice of using a timber plank to bridge an opening – material may be unsuitable to take a person's weight or damaged/weakened in previous use
- Operative too close to working plant

Operative No. 13
- Riding on a dumper (or any plant for that matter) without proper seat fitted is contravening safety regulations and is a sackable offence

Operative No. 14
- Signalman for crane driver should be positioned in full view of crane driver at all times, not to the side. Crane driver will need to read the hand signals looking straight ahead

Scaffold comments

- It could be argued that the type of scaffold shown in this example is the incorrect type to use as the full height of the structure is not shown. Scaffold is a **putlog** type (often referred to as a bricklayer's scaffold) and usually does not go higher than two storeys. A better, more substantial, scaffold would be an **independent** type in order to prevent a scaffold collapse
- Overhang of some ends of scaffold boards in excess of regulations, liable to cause lift of boards if trodden on with operative falling through
- Insufficient width of working platform for operative and materials
- Insufficient width of working platform for operative and materials
- Insufficient tying-in of the scaffold at openings to prevent the scaffold from falling away from the building
- Toeboards incomplete and having gaps in places
- Large gaps in abutting scaffold boards and joints
- Gaps/scaffold boards missing, allowing operatives to fall through
- Missing baseplate to some standards
- Sole plates missing to some standards
- Guard rails missing completely
- End of scaffold platform has no return guard rail or toeboard

Hoist comments

- No caging around the hoist sides to isolate/contain moving hoist and hoisting materials and to prevent operatives falling off the scaffold down the hoist opening
- No gates at ground level – first lift – second lift and third lift, etc.
- Plant restricting full access to hoist platform at ground level
- Hoist-operating machinery not caged in
- No landing stage for wheeling off materials on second or third lift

Trenching comments

- Insufficient struts used to hold each side of the trench apart to prevent a cave-in

- Planking to sides of trench need to be extended above ground level in order to prevent materials/tools from falling in on operatives

Plant in general

- No warning poles with danger bunting to prevent plant getting too near to overhead electricity supply lines
- No large timber fenders fixed to the ground to prevent the plant from getting too close to an open trench excavation

- Crane too close to excavation
- Wrong type of hook and sling used
- Electrical lead running through puddle – also trip hazard for other operatives

Materials generally (good housekeeping)

- Timber left littering the site – should be neatly stacked
- Pipes not prevented from rolling

Answers to selected self-assessment tasks

2.1
1. 1000
2. mm², m²
3. m²
4. kilo means 1000
5. 1000

2.2 kilonewton per millimetre squared; millimetre per metre per degree Celsius; watt per metre per degree Celsius.

2.4
1. (b)
2. (a)
3. (c)
4. (b)

2.6
1. $1057\,\text{kg/m}^2$. It is *not* sensible to write the answer as $1057.082\,452\,\text{kg/m}^3$ just because it is displayed on the calculator. Answers should always be rounded off to a sensible number of significant figures.
2. The density of the common brick turns out to be 10 times too small. This is because the volume had been calculated *incorrectly*. It should be $0.0014\,\text{m}^3$.
3. (d)

2.7
1. $1000\,\text{kg/m}^3$

2.11
1. Heat flow is into building
2. No heat flow
3. Heat flow is out of building

2.14
1. (i): $0.1\,\text{W/m}\,^\circ\text{C}$

2.15
1. 15 mm (approx.)
2. (c)
3. (c)
4. (iii) $0.15\,\text{mm/m}\,^\circ\text{C}$, (iv) polythene

2.16 Care should be taken with this question. The reader should find that the answer is 45% porosity.

2.20 2. (c)

2.21 (d)

2.22
1. Stress $= 1200\,\text{N/mm}^2$, strain $= 0.02$. Now check your final answer with the value in Table 2.9. Give two reasons why the two values do not agree exactly.
2. Approximately $96\,\text{kN/mm}^2$. This is very close to the elastic modulus of copper.

2.26
1. $55\,\text{N/mm}^2$
2. Steel is 10 times stronger in tension
3. (d)

2.30 No. Even with defect-free specimens, there will always be some variation in the strength due to slight differences between specimens.

2.31
1. $22\,032\,\text{mm}^2$, $50\,\text{kN/mm}^2$
2. $36.3\,\text{N/mm}^2$

2.33
1. $1040\,\text{kN}$
2. $46.2\,\text{N/mm}^2$

2.34 Model answer for task 2:

Definition: Compressive strength $= \dfrac{\text{maximum load}}{\text{area}}$

Rearrange this expression:
maximum load $=$ compressive strength $\times$ area
Given: compressive strength $= 2.0\,\text{N/mm}^2$ (from Table 2.13)
Calculate: area $= 305 \times 5000 = 1\,525\,000\,\text{mm}^2$
Finally, substitute values into formula for maximum load:
maximum load $= 2.0 \times 1\,525\,000 = 3\,050\,000\,\text{N}$
$= 3\,050\,\text{kN}$ (thousand N)
$= 3.05\,\text{MN}$ (million N)
(Note the different ways of expressing the answer.)

2.35 Strength in tension is much greater.

2.36
1. $7900\,\text{N}$ and $2500\,\text{N}$
2. $79\,\text{N/mm}^2$ and $25\,\text{N/mm}^2$
3. (a)

2.38 2. (d)

2.51
2. (d)
3. (c)
4. (e)
5. (a)
6. (c)
7. (a)
8. (b)

Index